Broadcasting
Britain

100 years
of the BBC

Robert Seatter

For opera lovers
13 November 1936: viewers are
treated to excerpts from *Mr
Pickwick* – the first time opera is
broadcast on British television.

Broadcasting Britain

100 years
of the BBC

Robert Seatter

Contents

The Art Deco front entrance
of Broadcasting House in
London's West End.

Introduction

Robert Seatter is an award-winning poet, performer, broadcaster, and arts professional. He has worked in broadcasting at the BBC for over twenty years in a range of roles, both in front of and behind the microphone/camera. He is Head of BBC History, telling the story of the Corporation at the heart of national life, as well as on the global stage.

How to tell the story of 100 years of the BBC?

It begins with innovation, as technology does its magic trick of transforming our lives. And so "this miraculousl toy" leaves the radio ham's shed and enters the living room. BBC radio voices fill our lives in the 1920s, 30s and 40s… Beatrice Harrison and her nightingale, King George V with his first Christmas message, John Snagge on the sports field, Tommy Handley helping us laugh at dictatorship, Vera Lynn's songs. The list goes on and on.

It's a shared experience, too. Suddenly, we have a new, extended BBC family. And flash forward a few decades, the *Manchester Chronicle* advises us to "Put an H-aerial up over your house, and you will be astonished to find how many friends you have in the street".

BBC television reflects our lives, and it also brings new worlds into them, challenges them, excites and soothes them in equal measure. There are wars and watershed moments – the Falklands to 9/11, there is the miracle of the moon landing, there are provocative Plays for Today, anarchic comedy and cosy sitcoms, as well as Eurovision hits and classical "firsts", conversations with gorillas and the sky at night, nonsense-speaking Flowerpot Men and terrifying Daleks.

How to tell the BBC story, and what indeed to leave out? The answer is so much. This book doesn't aim to be comprehensive: to do so would be foolhardy. It aims to gather a cluster of highlights from each year of the BBC's lifetime to paint a picture of the BBC at the heart of a constantly changing Britain and a wider world; and also show how BBC broadcasting is always in the "now", while trying to make sense of the "before" and the "then".

I hope you enjoy *Broadcasting Britain: 100 Years of the BBC*.

Robert Seatter

Foreword

Nick Robinson is a well-respected and familiar face and voice on television and radio with over 30 years' experience in reporting and broadcasting. He joined the flagship Radio 4 *Today* programme as a presenter in November 2015, after a decade as the BBC's Political Editor covering Prime Ministers Blair, Brown and Cameron. His *Election Notebook* was a *Sunday Times* bestseller.

If the BBC didn't exist would anyone think of inventing it? Would anyone take seriously the mission set out a hundred years ago by the first Director-General, John Reith "to bring the best of everything to the greatest number of homes"? Would they believe that it was possible, as he pledged, to give people "all the facts… presented to them in such a way that it is possible for them to make up their own minds"?

You can imagine the response to such bold claims in Parliament or in the pages of the newspapers or on radio phone-ins. "Who do these people think they are?" would be the cry.

In fact, you don't really need to imagine it as this is precisely what some say now. They suggest that, after a hundred years, the BBC should either accept its time is up or be drastically cut down to size. After all, they argue, what was right for an age when there was only one broadcaster is no longer right when there's almost an infinite number – whether direct competitors such as SKY and ITV, Netflix or Amazon Prime, YouTube or kids with nothing more than mobile phones posting on Instagram.

Yet as I drove from Kyiv to the Polish border a few days after the war in Ukraine began, the mere sight of those three letters – "BBC" – was enough to have our car waved through checkpoint after checkpoint. I doubt many there knew – I doubt many here know – the motto on the BBC coat of arms "Nation shall speak peace unto Nation." What they did know – and what most people who pay for the BBC know – is that the BBC is trusted in a way almost no other organization is.

Ah, say the critics. That's what they say abroad. That's what people used to say here, but not any more. Yet, why then did millions turn to the BBC for impartial news about the COVID pandemic – on TV, on the radio and online?

Now no one knows better than someone who's spent most of their adult life working at the BBC that it can and does get things wrong.

However, for all its flaws – the BBC and the people working for it still strive to fulfil the lofty ambitions of John Reith, instead of promoting a party, cause or set of beliefs held by the wealthy, the powerful and the well connected.

Yet there is another reason the BBC continues to be revered by so many. It is captured in this book. Every page of *Broadcasting Britain* is a reminder of a programme we loved; a presenter we trusted; a theme tune we can still hum or a shared national moment. As I read it, I recall watching Morecambe and Wise as my Dad laughed so much he cried. I remember the thrill of the *Match of the Day* theme and the joy of having our tea on a tray so we could watch Wimbledon. I remember the emotional power of *Boys from the Blackstuff* and the real-life drama of the *Ten O'Clock News* or the *Today* programme – which I'm lucky enough to present – during a national or international crisis.

You will have your own favourites – some from long ago, some, perhaps, from only last week… programmes that have shaped all our lives… programmes that help to explain why those three letters, "BBC", are as good as any passport when you're in a warzone.

Nick Robinson

Nick Robinson presents the *Today* programme on Radio 4, 2015.

The 1920s Radio Magic

Silence. Then radio waves come through the ether, and life is never the same again. Suddenly at the touch of a button can be heard the voice of the king, regular news, the repeating rhythm of the pips, dance music, daily prayers and guidance for shipping. It is… magic!

Experimenting with sound effects in the Savoy Hill studio, March 1927.

The British Broadcasting Company

"I went round to my next-door-neighbour's house, and I saw this round cone on the wall. And I went in to my mother and I said: "Mother, Mrs Buckle's wall is singing." "Don't talk so silly," she says. So I says, "There is, there's some music coming out of the wall." So she went to have a look and she said, "Oh, it's amazing…!"

A Yorkshire woman recalls the first time she heard the radio.

A classic crystal set for early radio listening.

In this year

18 October
Founding of the British Broadcasting Company Ltd, a consortium of leading wireless manufacturers.

14 November
The first broadcast.

2 December
John Reith appointed General Manager.

In the 1920s, there are no radio programmes as we would understand them today, merely a very basic form of wireless communication, the hobbyist preserve of boys and men. Cartoons of the day depict men in sheds surrounded by a paraphernalia of wires and flashing bulbs as they construct their own wirelesses, while wives in the kitchen ponder what their husbands are up to.

From a government perspective, there is concern that the commercial radio companies that are gradually emerging to serve a possible future audience may be a risk to national security, creating a "chaos of the ether", as in the US. They also see the potential of a lost income stream.

"Listening in"

So on 18 October 1922, the government brings together the six major wireless manufacturers, plus a few smaller companies, to create the British Broadcasting Company. Its first official broadcast follows from London on 14 November, opening with the immortal words: "2LO calling" – 2LO being the name given to the Company's radio transmitter.

Once liberated from the garden shed, radio – and the pastime of "listening in" as it was called – quickly becomes the passion of the decade. There is even a popular song specially composed in its honour:

"I'll B.B.C-in' you, it's the latest craze,
I'll B.B.C-in' you, it's quite the craze.
I'll B.B.C-in' you, that's what they say,
I'll B.B.C-in' you, most any day."

One radio "first" quickly follows another, as the BBC discovers what can be done, content wise, with "this miraculous toy", in the words of its first manager, John Reith. In little over a month in this year, the BBC chalks up the first general news bulletin; the first talks programme; the first *Children's Hour*; and the first religious address.
Silence no more.

John Reith

"What I was capable of compared with what I've achieved is pitiable."
John Reith

John Reith in 1926. As Director-General, he gives the BBC its mission: to inform, educate and entertain.

A towering figure, physically as well as metaphorically (he was 6 feet 6 inches tall), John Reith is the BBC's first General Manager, and the man who really shapes this new Company. Born in Stonehaven, Scotland, he is the son of a church minister and the youngest of seven children. As with many of the men who were there at the beginning of the BBC, he was marked by the devastation of World War I, which had ended just four years earlier, leaving him with a desperate desire to do something great and good in the world.

In 1922, he answers an ad in the newspaper for the post of General Manager at the BBC, although, as he confides to his diary, "I know nothing whatsoever about broadcasting". But then no one else really does, either. However, very quickly, Reith grasps the importance of the new medium, and writes his brilliant manifesto for it in his 1924 book *Broadcast Over Britain*. Here he creates both the templates for public service broadcasting in Britain and for the arms-length public corporation that is to follow. Becoming the BBC's first Director-General in 1927, Reith remains in post until 1938.

11

The First Radio Times

"Hullo Everyone! We will now give you the *Radio Times*. The good new times. The Bradshaw of Broadcasting. May you never be late for your favourite wave-train. Speed 186,000 miles per second; five-hour non-stops. Family season ticket: First Class, 10s per year."

The BBC's Director of Programmes, Arthur R. Burrows, writing in the first edition of the *Radio Times*.

The first colour edition of the *Radio Times*, ready for Christmas, 1923.

Radio is gaining a profile, but General Manager John Reith is keen to get the BBC's listings printed in newspapers in order to increase the popularity of programmes. However, there is resistance from the Newspaper Proprietors' Association (NPA), part of a growing conflict with the press of the day, who feel threatened by this new upstart "Radio". And so Reith decides to publish his own magazine – the *Radio Times*.

Rather than setting up an in-house publication, Reith seeks a publisher willing to undertake production on the basis of a profit share. Only one publisher is ready to take the risk, George Newnes Ltd, publishers of the frivolous popular weekly *Tit-Bits*.

Newnes is given editorial control and appoints Leonard Crocombe (editor of *Tit-Bits*) as the *Radio Times*' first editor. The BBC's role is merely to supply programme listings and information. Perhaps the arrangement also makes it easier to defend accusations that, through the *Radio Times*, the BBC is accepting advertising by the back door.

"The Official Organ"

There is no information on how the magazine's name is arrived at, but from the first edition, the subtitle is "The Official Organ" of the BBC. At 2 pence a copy, the initial print run of 250,000 soon runs out. Circulation quickly reaches 600,000, aided by new BBC stations across the UK, all of which have separate listings.

Colour arrives with the first Christmas edition, with a special full-colour cover and a higher price of 6 pence. By 1924, circulation is up to 750,000. The *Radio Times* remains one of the best-selling magazines in the UK.

The First Radio Licence

An annual licence fee of 10 shillings (50p) is first introduced under the Wireless Telegraphy Act of November 1923, to cover radio sets. At the time, labourers are earning around £2 12s (£2 .60) a week, so that gives some idea of the value of the Licence Fee. By the end of 1923, 200,000 licences have been issued and by 1928 this has risen to 2.5m. The payment of a universal Licence Fee is designed to formalize and build the UK radio industry. It's also there to create a funding stream to the government via the Post Office, which is officially responsible for this new industry, and the BBC (which initially receives 50 per cent of the revenue).

However, from early on, some newspapers champion those who refuse to pay, calling them "heroes of free enterprise". There are also concerns about the BBC as a monopoly broadcaster.

The Licence Fee will continue to be hotly debated over the decades, with questions around its imposition as a "broadcast tax", and concern, too, that governments can use its negotiation to control and coerce the BBC. As yet, however, no realistic alternative funding model has emerged.

An early radio licence, issued in June 1923.

Big Ben and the Pips

"Our bright idea was to let the world hear Big Ben… those chimes that link Britishers together the world over."

A. G. Dryland, the BBC engineer responsible for the initial recording of Big Ben

Behind the famous clock face of Big Ben, 1928.

All change for time in 1924. Overnight, the radio becomes an essential part of our national time keeping.

On New Year's Day 1923, the chimes of Big Ben are heard for the first time on the airwaves, heralding the New Year. The broadcast is so popular that in February, the chimes become a daily time signal, and a national institution is born, resonating to this day.

Initially, BBC engineers are not allowed inside the building so have to access the Clock Tower – which houses Big Ben – from the roof of the Palace of Westminster. As a result, their microphone picks up extraneous traffic noise as well as the chimes. However, before long, appropriate, low-sensitivity microphones are fitted right by the bells. During World War II the "bongs" of Big Ben are broadcast to occupied Europe and prove a great morale booster.

The Pips

Also this year, the pips arrive: six short electronically-generated beeps to indicate the precise time of the day. Invented by the Astronomer Royal Sir Frank Watson Dyson and the BBC's John Reith, the time signal was originally generated from the Royal Greenwich Observatory, but since 1990 the BBC has been responsible for it.

BBC Radio announcers and presenters do all they can to avoid "crashing" or speaking over the pips, but very occasionally things do go wrong and Radio 4 listeners respond immediately!

In this year

5 February
The first Greenwich time signal.

17 February
The Big Ben daily time signal.

23 April
George V delivers the first royal address to the nation.

19 May
The first "Nightingale Lady" broadcast.

13 October
Labour Prime Minister Ramsay MacDonald delivers the first election address.

Time signal apparatus in the Royal Greenwich Observatory.

The Cello and the Nightingale

"A glamour of *romance* has flashed across
the prosaic round of many a life."

John Reith, then General Manager of the BBC, on the imapact of this unique broadcast.

Beatrice Harrison (and dog), playing
the cello in her garden at dusk.

On 19 May, listeners hear for the first time an extraordinary duet live from a Surrey garden. It features the famous cellist Beatrice Harrison playing to a nightingale, and the bird – seemingly attracted to the sound of her cello – responding with its own song.

Harrison first became aware of the birds one summer evening when practising her cello in the garden. When this duet is repeated night after night, Harrison persuades the BBC that it should be broadcast. John Reith initially balks at the idea, complaining that the birds would be unpredictable prima donnas.

Would the birds sing?
It's a tense moment, as this is live broadcasting, as well as being one of the first BBC outside broadcasts. The BBC interrupts the Savoy Orphean Saturday evening performance to go to Harrison playing Elgar, Dvořák, and the Londonderry Air. No birds sing. Finally, 15 minutes before the end of the broadcast, the nightingales start chirping. Harrison plays and the nightingales, eventually, join in. (Some have since questioned whether a whistle may have been used to simulate the bird for the first recording; however, these assertions are unproven.) The public reaction is phenomenal and the experiment is repeated every spring for the next 12 years. Harrison and the nightingales become internationally renowned and she receives 50,000 fan letters. Some come from overseas, addressed simply to "The Nightingale Lady".

15

The First Broadcasts for Farmers

In this year

19 February
The first broadcast for farmers: *Farmer's Talks*.

27 July
The Daventry transmitter, the world's first Long Wave transmitting station, opens.

This week in February sees the announcement of new national radio programmes targeted at farmers and other rural workers.

The specific new broadcast of 19 February provides farmers with key information about market prices. It sits alongside other very specialized talks sprinkled across the airwaves, for example Ducks for Egg Production, Experiments on the Manuring of Roots in Dorset, and Phosphatic Manures for Spring Use.

Highly specialized though these programmes are, they develop a general listenership and become a recognized part of the BBC radio landscape. Much as still happens today, when programmes such as *Farming Today* (1960) serve the agricultural and political community, but also provide a glimpse of rural life for wistful urban dwellers.

The Daventry Transmitter

On this day, the world's first Long Wave transmitting station opens. The transmitter, known as 5XX, is positioned on Borough Hill near Daventry, Northamptonshire, to cover the maximum land area.

Before Daventry, the BBC's radio coverage across the UK is decidedly patchy. This new transmitter brings the total audience within listening distance to 94 per cent of the population. It makes the idea of a BBC nationwide radio service a reality.

Signposts of radio's future:
Daventry transmitter masts.

Radio is still new enough to be a source of wonder, and the opening this year is a big event. A specially commissioned poem by Alfred Noyes, is followed by speeches from the Postmaster-General and the Mayor of Daventry, introduced by Lord Gainford, BBC Chairman. Even Prime Minister Stanley Baldwin sends a message, which is published in the *Radio Times*:

"The opening of the Wireless Broadcasting Station at Daventry... will give no less than 20 million people the opportunity to receive both education and entertainment by means of cheap and simple apparatus; and I look upon Daventry as another milestone on the road to the social betterment of our people."

"Daventry calling… Wind and rain,
Against my voices light in vain.
The world through which my messages fare
Is not of the earth and not of the air."

"The Dane Tree" by Alfred Noyes, written for the inauguration
of the Daventry Transmitting Station.

Inside the Wireless Broadcasting Station at Daventry.

The General Strike

"The BBC fully realizes the gravity of its responsibility to all sections of the public, and will do its best to discharge it in the most impartial spirit that circumstances permit."

BBC news bulletin, 10.00 a.m., 4 May.

For nine days in May 1926 the nation's industry was brought to a standstill by the General Strike. This was a walkout by millions of dissatisfied British workers, particularly miners, in response to reduced pay and increased hours, following an overall decline in profits owing to the strength of the pound. The workers' strike slogan was, "Not a penny off the pay, not a minute off the day." Most newspapers stopped printing, so there were few means of communication available.

It was both a baptism of fire and a challenge for the BBC, especially given government restrictions, defined at the time of the Company's creation, over "controversial broadcasts" – basically anything of a political nature.

A food convoy passes through Hyde Park accompanied by armoured cars during the General Strike.

In this year

4–12 May
The General Strike
divides the nation.

26 September
First broadcast of
The Epilogue, a moment
of reflection marking
the end of the day.

December
Singer Edna Thomas is the
first known Black lesbian
female to perform
on the BBC.

"The sensation of a general strike, which stops the press,
centres round the headphones of the wireless set…"

Diary entry of writer Beatrice Webb reflecting on the importance of
BBC radio as a lifeline during the General Strike.

What role should the British Broadcasting Company play in these troubling times? The Conservative government publishes its own *British Gazette* newspaper, in effect a government propaganda sheet, launched and edited by the Chancellor of the Exchequer, Winston Churchill, who furiously opposed the strike: "It is a conflict which, if it is fought out to a conclusion can only end in the overthrow of parliamentary government or its decisive victory." Churchill could see that radio was an immediate and versatile medium of communication in the rapidly unfolding chaos of the strike, and he lobbied Prime Minister Stanley Baldwin to commandeer the BBC.

BBC General Manager John Reith lobbied back, arguing, in a series of exchanges with Baldwin, that such a move would destroy the Company's reputation for independence and impartiality. It remains a tribute to Reith's acumen and persuasiveness that he, and not the political heavyweight Churchill, won the day.

And so Baldwin ruled, just before the strike ended, that the BBC should remain independent – "Monstrous," Churchill called it. Reith argued later that Conservative, Labour and trade union perspectives in the crisis had been reported even-handedly; however the Labour Party and the TUC did not forgive the BBC for refusing airtime to their representatives – including Leader of the Opposition Ramsay MacDonald.

The importance of impartiality

Uncertain though the conclusions were around the exact role of the BBC during the General Strike, the importance of the organization as an unpartisan voice, independent from the government of the day was established. It led directly to its transition from a Company to a Corporation. This was discussed in detail at the end of 1926 and officially inaugurated with a Royal Charter, complete with a Great Seal, on 1 January 1927.

The BBC is defined by a Royal Charter
with its impressive seal.

19

The First Prom Concert

"With the wholehearted support of the wonderful medium of broadcasting,
I feel I am at last on the threshold of realising my lifelong ambition of truly
democratizing the message of music and making its beneficent effect universal."

Sir Henry Wood on the first BBC Prom.

Sir Henry Wood, the first conductor
of the Proms.

Founded in 1895 by impresario Robert Newman and under the inspired conductor's baton of Henry Wood, the "Promenade" concerts – cheaply ticketed standing concerts – were deliberately created to make classical music more accessible to everyone. However, by the mid 1920s, the organization was in dire financial straits and audience numbers were dropping, so, in 1927 the BBC takes them on, despite concern that radio might actually ruin the concert world.

Instead, the opposite happens, and the BBC Proms amplify the reach of classical music, reinforcing Wood's key message of the universal and democratic nature of music.

From Elgar to Sibelius

Relayed from the Queen's Hall, the first Prom features some English standards, along with a mix of crowd-pleasing classics. From Elgar's *Cockaigne Overture* to Boccherini's *Minuet in* A for strings, Liszt's *Hungarian Rhapsody No. 2* and Rossini's *William Tell Overture*, along with excerpts from *Coppélia*, "Elizabeth's Greeting" from *Tannhäuser*, and other songs and orchestral pieces

20

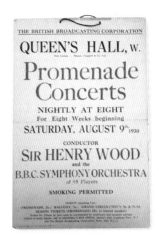

A Proms poster for the year 1930.

In this year

1 January
The British Broadcasting Corporation is created by Royal Charter; Sir John Reith is the first Director-General.

15 January
The first rugby commentary, followed later that year by the FA Cup, test cricket, and Wimbledon tennis.

13 August
The first promenade concert broadcast.

by Stanford, Sibelius, Quilter, Parry and Schubert.

For the first three years, the concerts are given by "Sir Henry Wood and his Symphony Orchestra", until the BBC Symphony Orchestra is formed in 1930, creating a strong orchestral link that endures to this day.

Following the 1941 bombing of the Queen's Hall during the Blitz, in 1947, the Proms relocates to the Albert Hall, where what is now the largest festival of classical music in the world still takes place.

The Proms diversify over the years with musical genres now ranging from jazz and folk to light entertainment and musicals. In addition, the BBC Proms takes music out of the concert hall and into schools, parks, squares, shopping centres and car parks.

The Royal Albert Hall has been the home of the BBC Proms since 1947.

21

The BBC Dance Orchestra

In this year

12 March
The first broadcast by the
BBC Dance Orchestra,
leader Jack Payne.

25 April
The first budget broadcast.

Jack Payne and the BBC Dance Orchestra light up the airwaves in Spring 1928, and soon have a huge following.

The band has already proved its popularity as The Cecilians, occasionally broadcasting from the Hotel Cecil in London's West End. Then, given the title of Director of the BBC Dance Orchestra, Payne moves his 10-piece band into the BBC studio at Savoy Hill, where they are a regular and popular radio feature. His signature foxtrot tune, "Say it with Music", soon becomes a hit.

The *BBC Handbook* of 1929 will acknowledge the importance of dance music on the radio, calling it rather grandiosely "the voice of something very typical of ourselves and of this post-war age". Radio dancing lessons are all the rage, and the listeners' appetite for dance music is huge. As a testament to that, the BBC Dance Orchestra is soon receiving 10,000 letters a week.

The popular Jack Payne and the BBC Dance Orchestra at Savoy Hill.

Payne will leave the BBC in 1932 to be replaced by Henry Hall, who continues the success with another line-up. Hall is followed by Billy Ternent and then Stanley Black, until 1952.

The orchestra provides the music for many hit programmes, including the radio comedy shows *Much-Binding-in-the-Marsh* (1944), *Ray's a Laugh* (1949) and *The Goon Show* (1951).

Henry Hall's guest night, with famous singer/composer Ivor Novello, 1934.

The First Budget Broadcast

In the early years of the BBC, there is a strict policy of avoiding broadcasts on controversial political issues. This relaxes slightly post 1927's Royal Charter, and in April 1928 Winston Churchill, then Chancellor of the Exchequer and never a particular friend of the BBC, makes a brief radio address on his forthcoming budget, the first time such a topic has been covered in any detail on the airwaves.

Director-General John Reith accompanies Churchill to the studio at Savoy Hill, and later writes in his diary: "He delivered a good defence of the budget, supposed to be non-controversial, but it was not."

The Labour Party's Ramsay MacDonald protests immediately, and after this incident, the political parties struggle to agree how to work with the BBC on political on-air discourse.

Winston Churchill at the BBC microphone.

The Week in Parliament

> "Broadcasting is clearly rediscovering the spoken language, the impermanent but living tongue, as distinct from the permanent but silent print."
>
> *Broadcasting* by Hilda Matheson, 1933.

**Hilda Matheson
(1888–1940)**

Hilda Matheson meets John Reith while working for Lady Astor, the first female MP in the UK. She quickly rises to become the first ever Director of Talks and also establishes the first radio News department. At this point, she is the most senior woman in the BBC. Her work in the Talks department transforms its output, making it a more authentic means of natural communication as well as focusing it on a wider range of listeners. Matheson resigns from the BBC in 1931, following a clash with Reith over his rigid control of content.

The world is changing. In 1928, women over the age of 21 were given the vote, making them equal to men – in this respect, at least – for the first time. And so the 1929 General Election becomes known as the "Flapper Election" as young women voice their political opinions for the first time, and 14 women are elected as MPs.

The BBC responds to this, in the person of the extraordinary Hilda Matheson, who decides to create *The Week in Parliament*, a platform for female MPs to broadcast short talks in order to help women familiarize themselves with the political process. The first such speaker is Margaret Bondfield, who has recently become the first-ever British female cabinet minister, taking on the post of Minister of Labour a few months earlier in June.

Writing to Lady Astor, her ex-employer, in November, Hilda Matheson explains the aim and audience of the programme: "A new experiment" is planned for 10.45 a.m., "the time when we find most busy working women can listen best, when they have their cup of tea".

New broadcasts bring the workings of Parliament to life.

Children's Hour

In this year

16 January
The Listener weekly magazine launches, providing in-depth insight into BBC programming. It runs until 3 January 1991.

19 July
The *Toytown* plays

6 November
The Week in Parliament (later renamed *The Week in Westminster*).

Children are at the centre of BBC programming from the very start. Launched in 1922, *Children's Hour* introduces the Uncle and Auntie presenters – Uncle Arthur, Uncle Caractacus, Uncle Jeff, Uncle Rex, Auntie Sophie, Auntie Phyllis... They are almost certainly the origin of the nomenclature "Auntie BBC". In 1926, when the BBC removes the Auntie/Uncle presenters, there is an outcry from its young audience, and back they come a year later. When the BBC finally cancels *Children's Hour* in 1964, 60 MPs sign a letter of protest!

The show features radio's first commissioned drama, first orchestral piece, first zoo visit, and first animal sound – Uncle Jeff's dog "The Grand Howl"! It also includes popular features and stories, none more so than *Toytown*, which airs on 19 July and lasts until 1963.

Larry the Lamb and friends...

Narrated by Uncle Mac (Derek McCulloch), also the voice of lead character Larry the Lamb, *Toytown* introduces children to the soon-to-be-loved characters of Dennis the Dachsund, the Mayor, Ernest the Policeman, Mr Growser, The Inventor, The Magician, Dennis the Artist, Captain Higgins, Mrs Goose and Letitia Lamb.

The *Toytown* stories came from the pen of S.G. Hulme Beaman, who wrote and illustrated the book *Tales from Toytown*, as well as

Popular *Toytown* characters make their television debut in *Puppet Theatre*, 1956.

carving the characters in wood. "Elizabeth" from *Children's Hour* (the presenter May Jenkins) discovers the book and prompts its adaptation to radio. Despite Hulme Beaman's early death aged 45 in 1932, *Toytown* persists, moving to television in 1956.

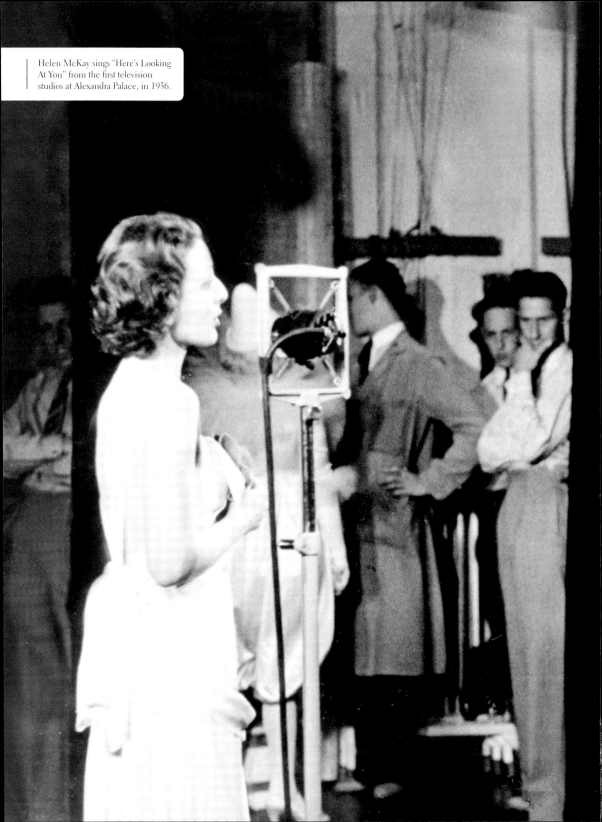

Helen McKay sings "Here's Looking At You" from the first television studios at Alexandra Palace, in 1936.

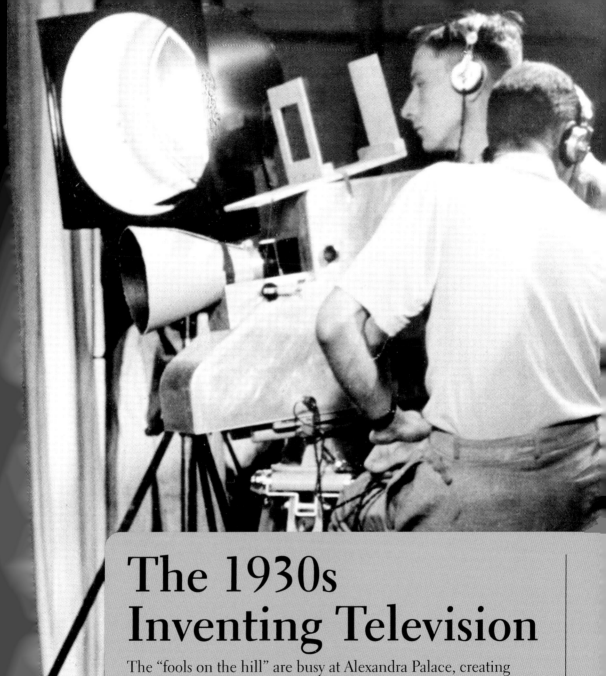

The 1930s
Inventing Television

The "fools on the hill" are busy at Alexandra Palace, creating a whole new medium. A television set costs the price of a car, its programmes are viewed only on a tiny screen, and it is perhaps a threat to the civilized world. Human beings may never leave their houses again…

The Man With the Flower in His Mouth

Producer Lance Sieveking (top left) directs actors George Inns, Gladys Young, Earl Grey, C. Denis Freeman, Lionel Millard and Mary Eversley.

Slowly, slowly, television is beginning, as the Scottish inventor, John Logie Baird, starts his public experiments, attempting to engage the wider public in this new innovation.

Caught up in this mood of excitement, the director of BBC Radio Drama Val Gielgud (brother of the famous actor John Gielgud) instigates the first experimental play, *The Man With the Flower in His Mouth* by the Italian playwright Luigi Pirandello.

He has a keen producer called Lance Sieveking and a company of enthusiastic actors, all of whom throw themselves with zest and brio into the project. But it is a taxing process – only one figure can be projected at a time and that figure is scarcely able to move.

The 14 July broadcast remains an eccentric one-off, and the BBC management determines not to over-commit to television… yet.

The BBC Symphony Orchestra

"One of the finest orchestras in the world."

The conductor Arturo Toscanini to Adran Boult on the BBC Symphony Orchestra in 1933

In this year

18 April
"No news day"
BBC news finds nothing significant to report, so music is played instead.

14 July
The Man With the Flower in His Mouth by Luigi Pirandello.

22 October
The first broadcast by the BBC Symphony Orchestra.

The first broadcast performance by the new BBC Symphony Orchestra is under the baton of famed conductor Adrian Boult in a concert relayed from the Queen's Hall, London.

It is an eclectic programme, featuring works by Wagner, Brahms, Ravel, and Saint-Saëns' Cello Concerto No. 1.

The BBC Symphony Orchestra has quickly assembled a full-time complement of 114 players, ready to tackle a wide-ranging repertoire, and soon its Sunday Evening Concerts will attract the biggest audience of the week and bring classical music live into living rooms up and down the land.

From the start the orchestra is committed to new music, premiering works by composers such as Ravel, Schoenberg and Holst. In fact, since its creation the orchestra has performed over 1,000 specially commissioned works, by composers from the minimalist John Adams to Master of the Queen's Music Judith Weir.

Today, the BBC Symphony Orchestra continues to be the mainstay of the Proms, and gives concerts at home and abroad. All concerts are broadcast on Radio 3.

The BBC Symphony Orchestra conducted by Sir Adrian Boult, at the Queens Hall.

Programmes for Gardeners

"Mr Middleton says: The more you grow the less you buy!"

World War II uses the wise words of C.H. Middleton to encourage gardeners to grow their own produce, as true now as then.

C. H Middleton, the popular voice of the BBC's first gardening programmes.

Gardening begins its perennial broadcast popularity in May this year, as the first-ever programme for gardeners hits the airwaves. It is structured as a series of scripted talks – all broadcasts are scripted in these days – but the choice of C. H. Middleton as its first presenter is inspired, as he manages an on-air style that is both informed and knowledgeable, yet also easy and natural.

He quickly proves to be a great success. *The Week in the Garden* becomes *In Your Garden*, moving to a regular Sunday slot and attracting a regular three and a half million listeners.

Middleton will soon transfer his gardening presenter skills to the new medium of television too, when the fledging service begins in 1936. A special garden is even created at Alexandra Palace, so that outside broadcasts can show live demonstrations.

During World War II, Middleton becomes the figure behind the government's Dig for Victory campaign, encouraging people on the Home Front to grow their own food. Mr Middleton dies in 1945, but *In Your Garden* will continue until 1970. And of course gardening programmes still thrive on both BBC radio and TV, including *Gardeners' Question Time* (1947) and *Gardeners' World* (1968).

C. H. Middleton in the garden at Alexandra Palace, 1937.

Innovation: The Blattnerphone

9 May
Mr Middleton presents the first gardening programme.

2 June
The Epsom Derby is the first live TV outside broadcast.

The BBC has no viable means of recording sound until this year when they purchase the extraordinarily-named Blattnerphone, designed by early British filmmaker Louis Blattner. This huge device uses 6 mm (quarter-inch) steel tape to record a very basic audio signal – good enough for voice, but not for music.

Spools are large and heavy, and editing is done by soldering the tape. But the speed at which the machine runs, 1.5 metres (5 feet) per second, means recording is hazardous for the operator – a break in the tape can result in razor-edged steel flying around the studio.

And at the end of the decade, in 1939, the Blattnerphone is famously used to record Prime Minister Neville Chamberlain's announcement of the outbreak of World War II.

The impressive spools of the innovative Blattnerphone recording machine.

Broadcasting House Opens

The brain centre of modern civilization... the world's new voice.

The Daily Express, 1931

The Art Deco splendour of the Concert Hall, the biggest public space in Broadcasting House.

Designed by architect George Val Myer and opening with a great fanfare in May of 1932, Broadcasting House is the BBC's first bespoke building – and the first ever purpose-built broadcast centre in the UK. And what a change it makes!

From 1922 to 1932, the BBC was mainly based at Savoy Hill, in what was then the Institution of Electrical Engineers, next to London's famous Savoy Hotel. It was cosy and convivial, but far from ideal in broadcasting terms. But all that is transformed by the opening of Broadcasting House. It is modern in every way, with cutting-edge studios equipped with state-of-the art technology in the building's central tower, as well as speedy lifts and effective ventilation.

A vision for the future

"Temple of the arts and muses"– these are the first words of the gilded inscription on the ceiling arch of Broadcasting House's main reception. They welcome workers and visitors alike to a palace of art and creativity: such is the vision for BBC Radio in these early decades.

The modernist elegance of Broadcasting House reception.

The streamlined West façade of Broadcasting House.

The building has a classic Art Deco design. With its accentuated front section bearing a clock tower and aerial mast, it is often compared to a streamlined 1930s ocean liner. It is also adorned with striking statues of Ariel and Prospero on the front façade of the building by Eric Gill. A second statue called The Sower stands centrally in the main reception. Hand in pouch, he is about to scatter seed over and across the fields, just as the BBC stands at the centre of the nation diffusing radio broadcasts to anyone who wishes to listen. The statue is a symbol of universality, as important now as then…

Reinvention

Over the decades, various extensions will be added. Then, from 2000 to 2013, it undergoes a major architectural redesign by the architect Sir Richard MacCormac. He goes right back to the original 1932 building, adding to it a harmonious Portland stone extension and a new public space.

The reinvented Broadcasting House is now the centre for national and international news, as well as the home of core radio and television commissioning and the BBC's HQ.

Broadcasting House
Clockwise from top:
Streamlined studio 80A for
musical performance; constructing
the iconic radio mast, 1931; the
steel structure of the building rises
up, 1930; the finished building,
1933; the new, extended
Broadcasting House, with the spire
of All Souls church, 2012.

The Royal Christmas Message

A human voice – intimate and personal – speaking to them in their own living rooms, speaking to them from a box on the table between the sewing machine and the mug.

George V's biographer Harold Nicholson describing the popular impact of George V's broadcasts.

George V delivers the first broadcast Christmas Message.

"Through one of the marvels of modern Science, I am enabled, this Christmas Day, to speak to all my peoples throughout the Empire." So begins the first Christmas Day message by a British monarch, when George V launches his live broadcast from Sandringham. The speech is written for the King by Rudyard Kipling, famed author of The Jungle Book and the poem 'If'. In the speech, Kipling celebrates the power of the wireless to unite not just the family gathered around their radios at home and throughout the nations of the United Kingdom, but, momentously, all the peoples of the Empire (as it was then).

Indeed, the King's speech comes at the end of a special programme, *All Over the World*, where royal subjects from Canada, New Zealand, Australia, South Africa and Gibraltar send Christmas greetings to their far-flung relatives at home. And the time of 3.00 p.m. is chosen as the best time to ensure the King's broadcast is heard in the largest number of countries across the Empire, creating a new sense of radio as a means of international communication and dialogue.

A royal Christmas

The idea for the royal broadcast comes from John Reith himself, who actually suggested it five years earlier as he could see then how potent the combination of two British institutions – the monarchy

George VI broadcasts the royal Christmas Message in 1944, inspiring the nation during wartime.

The technical set up for the Christmas Message with Her Majesty the Queen, 1960.

The Christmas Message with Her Majesty the Queen, from Sandringham, 1994.

and the BBC – might be. The King, however, is reluctant, and it takes the persuasive words of Prime Minister Ramsay Macdonald to convince him to do it. The public's reaction to the broadcast is hugely favourable, however, and the King himself subsequently declares that he is "very pleased and much moved" by the response.

And so a national institution is born, and the Christmas message from the monarch becomes a regular broadcast fixture. Later, radio will give way to television in 1957, and in 1997 it will be made for the first time by ITV rather than the BBC. This follows the revelatory *Panorama* interview by Princess Diana – though the Palace refutes this as the reason behind the change, declaring the intention is rather to work in partnership with a range of broadcasters.

The First Female Radio Announcer

"We foresee panic among the horse-hair armchairs," writes the *Radio Times*, imagining outcry from male listeners on hearing a woman announcer; but surprisingly many women complain about her.

On 28 July, Sheila Borrett becomes the first female announcer on BBC Radio. There is a high degree of audience excitement around the appointment, for two reasons – the other announcers are all men and also, they are anonymous. Anonymity is felt to be significant during this period, in order to ensure that the personality of the presenter in no way distracts from the message being conveyed.

Sheila Borrett is an experienced radio actor, and is called the "woman with the golden voice" by the press of the day. Sadly, she only lasts three months as the BBC receives over 10,000 letters of complaint. Happily, the much-maligned Sheila returns to acting after her brief announcing career ends.

In Town Tonight

"Once more we stop the mighty roar of London's traffic and, from the great crowds, we bring you some of the interesting people who have come by land, sea and air to be *In Town Tonight*."

Opening credits from *In Town Tonight*.

Mr and Mrs Gary Cooper, just two of the stars encountered *In Town Tonight* during 1938.

Variety programmes rapidly grow in popularity during the 1930s, and none proves more popular with audiences than *In Town Tonight*, in effect the prototype of the modern chat show.

The programme is introduced by Eric Coates' stirring signature tune "Knightsbridge March", mixed with the hurly burly of a London soundscape; suddenly the music is halted by the anonymous announcer declaring: "Once more we stop the mighty roar of London's traffic…"

In this year

21 April
The first TV revue,
Looking In.

28 July:
Sheila Borrett is the first
female BBC announcer.

18 November:
The first broadcast of
In Town Tonight.

And then the show begins, bringing to the microphone at the same time each week a great medley of characters who either live in or are visiting the capital.

Meeting the stars

And what guests turn up – film stars of the day, such as Gary Cooper, Errol Flynn and Rita Hayworth, as well as local characters and, later, presenter Brian Johnston performing madcap stunts, from riding a circus horse to lying under a passing train. The show has a huge following, and from 1954 until 1960 runs on both television and radio. Its popularity right across the UK is borne out by this enthusiastic comment from the Regional Programme Director in Scotland, who reports that, for his listeners, "It is the very elixir of life to be transported on Saturdays to [London], the hub of their heaven!"

Brazilian star Carmen Miranda
hits the town in 1948.

Innovation: the Droitwich Transmitter and the Classic BBC Mic

"I heard it on the BBC, it must be true."

The author George Orwell, who works at the BBC's Empire Service in the 1940s, describes his sense of confidence in the news he receives via the microphone of the BBC.

One of the transmitter's output valves is carefully wheeled into position.

The Daventry transmitter of 1925 has improved radio coverage across the UK. But the opening of the Droitwich transmitter in Wychwold, just outside Droitwich, Worcestershire, takes it to the next level. This is the most powerful transmitter permitted under international regulations, and, finally, more or less universal broadcast coverage is achieved across the UK – though transmitters in Scotland and Northern Ireland still function nationally.

The "top men"

The Droitwich transmitter opens in October 1934, broadcasting the National Programme on long wave and a Regional Programme on medium wave.

A film taken at the time shows the dangerous erection of the mast, at a height of 216 metres (700 feet), 61 metres (200 feet) taller than any previous transmitter. This massive structure is put in place not by a crane, but by four "top men" who, after a perilous half-hour climb each day, work suspended on a single, small, wooden platform at the mercy of buffeting winds.

Radio sets made during this year are marked with "Droitwich" on their dial.

Broadcasting across the nation: engineers in action at the Droitwich Transmitting Station.

In this year

7 October:
The Droitwich high-power transmitter replaces the Daventry transmitter.

A BBC icon

The BBC Marconi Type A ribbon microphone, made by Marconi's Wireless Company of Chelmsford, Essex, England is highly significant for a number of reasons. Firstly and most importantly, it enables the BBC to improve the sound quality of its broadcasts.

A ribbon microphone – called so because a thin metal ribbon inside it responds to sound – is particularly good in studio situations. Also, the double-sided design accepts sound from front and back, but not from the side, and is especially suited to the human voice.

Buying British

Initially, the BBC has its eye on an American version of this microphone, developed by the RCA Corporation. However, this proves too expensive, so the BBC works with the Marconi company, then one of the biggest UK-based technology companies, to develop their own version. This becomes a classic BBC methodology – to take the best market product and adapt it to their specific purposes.

The second reason for the mic's significance is that it quickly becomes a symbol of the BBC: its voice, its trust, its national – and increasingly global – significance. Over the years, countless politicians, singers, actors, thinkers and speakers are photographed behind its iconic shape, casting their voices out to the wider world.

The classic BBC microphone, which becomes a global symbol of the BBC's "voice".

Actor Laurence Olivier at the mic in 1943.

41

Alistair Cooke and the USA

"I prefer radio to TV because the pictures are better."

Alistair Cooke

The urbane Mr Cooke broadcasting
in 1946.

American Half Hour is the show that
begins broadcaster Alistair Cooke's
famous relationship with America.
"Mr Cooke will introduce us to the
everyday America, about which we
seldom hear. With him always in these
half-hours will be a small changing
army of resident and visiting
Americans" says the *Radio Times* of
the day. And the US Ambassador to
the UK, Robert Worth Bingham,
guests on the opening programme.

Adopting America

Cooke's radio career began a year
earlier as a cinema critic. Despite his
urbane, transatlantic accent, he was
born in Salford, Lancashire, and
educated at Blackpool Grammar
School and Cambridge University.

Having spent two years in the
United States on a fellowship,
immersing himself in American
culture, he brings that insight to his
writing and broadcasts. Eventually,
he will move to the US and, in 1941,
become an American citizen.

American Half Hour develops
Cooke's special talent for writing and
easy, informed delivery. It paves the
way for his future broadcasting
triumph, *Letter from America*.

Letter from America

The world's longest-running speech radio programme, *Letter From America* will begin on 24 March 1946. The initial agreement is for a mere 13-week series, but the programme proves such a hit with listeners that it runs for 58 years, only ending with Cooke's death.

"I had to offer", writes Cooke in the *Radio Times*, "a direct impression of anything of America that took my fancy. Not a diatribe, not a composed essay, but the first impression of an accident, a person, a landscape on the nervous emulsion of A. Cooke".

Over the course of 11 US presidencies, Cooke's *Letter From America* ranges from typically wry and acute observation of social mores to reports on world-shaking events, such as the assassination of John F. Kennedy in November 1963 and the 9/11 terrorist attacks. In 1972, Cooke also makes the critically acclaimed television series *America*, which is an international success.

Accolades pour in following Cooke's death in 2004, underlining the impact of his broadcasts in defining and maintaining the "special relationship" between the USA and Britain.

Alistair Cooke transfers his life-long
fascination for the USA to television
in his 1972 series *America*.

Innovation: the First High-Definition Television Service

"A mighty maze of mystic, magic rays
Is all about us in the blue..."

"Television", lyrics by James Dyrenforth, music by Kenneth Leslie-Smith, and sung by
musical comedy star Adele Dixon on the opening day.

Lead engineer Douglas Birkinshaw tests the latest camera equipment in the grounds of Alexandra Palace.

Television arrives! On 2 November, the BBC launches the world's first high-definition television service from Alexandra Palace in North London. Its transmitter mast quickly becomes the iconic symbol of Television and all it has to offer.

However, initially Director-General John Reith is uninterested in television, until eventually he succumbs to government pressure – officials are aware that Nazi Germany is developing television for propaganda use – and commissions Director of Television Gerald Cock to set up an experimental service to test the mechanical system devised by Scottish engineer John Logie Baird and an electronic system created by Electric and Music Industries (EMI).

"The fools on the hill" – as the television team is called by Broadcasting House management – throw themselves into the endeavour. But after a couple of months in 1937, it becomes clear that the EMI system is much the better, and Baird's dream ends, though for most people his is the name associated with pioneering television.

Various sites are considered for the development of television: Highgate, Hampstead, Shooters Hill, Crystal Palace. Eventually, Alexandra Palace – the Victorian People's Palace created in 1873, but accident-prone and regularly burned down – is chosen. This is primarily because of its height: 107 metres (350 feet) above sea level and so easier for signal reception. Moreover, the building is available immediately, though in poor condition: cold, drafty and derelict.

The first television schedule

So what did audiences see on this first day? The schedule begins with speeches from the government and BBC, followed by a newsreel, British Movietone News. Adele Dixon then sings the specially composed "Television" theme tune, leading into dance and comedy from the Black American duo Buck & Bubbles. After that, *Picture Page* introduces a medley of guests requested by viewers, from the aviator Jim Mollison to the tennis star Kay Stammers.

Television flourishes until 1939. It closes down on the outbreak of World War II as the Alexandra Palace mast is an obvious target for enemy bombers. However, the mast continues to be used in secret by radar engineers to deflect enemy aircraft.

Edward VIII's Abdication Broadcast

"At long last I am able to say a few words of my own…"

Edward VIII begins his famous broadcast.

Edward VIII makes his life-changing broadcast from Windsor Castle.

For the first and only time, a British monarch decides to abdicate, and does so over the BBC airwaves. The broadcast is hugely controversial, and comes after months of agonized debate between the monarchy and government, as well as growing press anticipation.

Finally, Edward VIII decides that he feels unable to remain king unless he is permitted to marry the American divorcée Wallis Simpson. That he decides to tell the nation and the world of his decision over the airwaves is highly significant. The broadcast goes straight into people's homes conveying a sense of directness and immediacy; and the charged rhetoric of Edward's address gives it a uniquely personal note – he was never so eloquent ever again. People weep as they hear the news.

It is the most-listened-to broadcast of the decade with an estimated audience of 10 million in the UK, and purportedly of 150 million around the world.

The Coronation of George VI

> "We had a very serious fault and suddenly everything went dead… the only thing to do was to give it an almighty biff. And it worked!"
>
> Tony Bridgewater, engineer in charge of the outside broadcast, recalls the moment the new TV equipment almost failed, just before the appearance of the royal procession.

The newly crowned George VI and his wife Queen Elizabeth greet the public from the balcony of Buckingham Palace.

The Coronation of King George VI and Queen Elizabeth in 1937 is the big event of the early television service and the first true outside broadcast using a mobile control van. The press declares it "the supreme triumph of television to date", and the flickering images of the king smiling as his carriage passes by the cameras capture the imagination of the viewers – small in number though they still are.

In technology terms, the equipment is new and very sparse. The BBC sets up three cameras – half the total number it owns – each side of Apsley Gate. Frederick Grisewood provides commentary as the royal procession approaches through Hyde Park and passes through the gateway. A control van is parked nearby, with a second van on standby with a wireless link to Alexandra Palace, in case any of the numerous cables should fail.

Broadcasting pageantry

The BBC reports that the Coronation is seen by 10,000 people. But more important than this number is the initiation of the BBC's role as the principal media observer and interpreter of royal pageantry.

To come post-war will be the Coronation of Queen Elizabeth II and a panoply of spectacular royal weddings, distributed globally as stunning spectacles of British tradition and style.

Television cameras capture the progress of the royal carriage for the first time.

The First Television Coverage of Wimbledon

"Wimbledon is a special place for me in so many ways, and I feel privileged to have been such a big part of it over the years."

Tennis player and BBC commentator Sue Barker.

The BBC's partnership with Wimbledon has transformed the popularity of tennis. Here, TV cameras cover the action on Centre Court in 2014.

In this year

12 May
The coronation of George VI marks the BBC's first use of an outside-broadcast van.

21 June
The Wimbledon tennis tournament is shown on TV.

A month later, the Outside Broadcasts team chalks up another first, as they cover the tennis championships at Wimbledon. Interest in the sport is currently at its peak, following a string of earlier victories by the British player Fred Perry. Unfortunately, this year Perry does not triumph in the Men's Final, which is won by the American Don Budge.

As an outside broadcast, Wimbledon is a real challenge. Microphones have to be positioned so as to pick up the sound, yet be protected from the elements and out of vision. Likewise the cameras need to capture activity in a limited space without obscuring audience visibility. The *Radio Times* of the day underlines the innovation of the technology, and the unique ability of the TV cameras to capture events as they happen.

The broadcasts from the Centre Court feature commentary by Freddie Grisewood and John Snagge, by now extremely popular and familiar on-air voices.

The BBC's links with Wimbledon actually extend back to 1927, when the first radio commentary was broadcast. Radio – and, increasingly, television – magnify the popularity of the tournament, and foster its growing global appeal. Wimbledon coverage is also used to promote the take-up of colour television when that arrives in 1967.

The BBC's coverage of Wimbledon remains the Corporation's longest-running sports partnership.

The First Foreign-Language Broadcast

"Britain's greatest gift to the world in the 20th century."

UN Secretary-General Kofi Annan on the impact of the BBC's World Service.

In this year

3 January
The first foreign-language service begins broadcasting, to the Middle East in Arabic.

5 January
Comedy show *Band Waggon* debuts, starring Arthur Askey and Richard Murdoch.

27 January
Douglas Byng is the first drag artist to appear on BBC television, in an entertainment career that spans nearly 70 years.

6 April
The Oxford and Cambridge boat race is shown live on TV.

Director-General John Reith introduces the BBC's first foreign language service – in Arabic – on 3 January, followed by the voice of announcer Ahmad Kamal Sourour Effendi, recruited from the Egyptian State Broadcasting Service in Cairo.

Content creation for the new service is not easy and there are clashes over what news items should be included to engage the complex interests and sensibilities of the Middle East. In addition, the Foreign Office is watchful. But the BBC is clear from the outset that it means to create a foreign-language broadcast model similar to that of the domestic BBC – independent from government interference.

The instigation to broaden the BBC's Empire Service into foreign-language coverage derives from the increasingly fraught overseas political situation. War is coming, and the Italian government under Mussolini is using a powerful transmitter to broadcast anti-British propaganda to the Middle East. The BBC decides to respond with programming that is more measured and balanced. The next few years will see a huge expansion of foreign-language output.

Ahmad Effendi's appointment makes the new Arabic service popular overnight as he is one of the best-loved presenters in the Arabic-speaking world.

In 1965, the BBC's Overseas Service is renamed the World Service. It continues to broadcast to nearly 300 million weekly listeners and viewers around the globe.

During World War II the BBC grows from eight to more than 40 language services by 1945. It also sees the development of the Empire Service as a lifeline news broadcaster, especially for occupied Europe where the broadcasts of General De Gaulle launch the Resistance. The Cold War years are challenging for the Service. It is blocked in many countries, and World Service journalists are often personally targeted, most notoriously in the case of Bulgarian correspondent Georgi Markov, murdered by a poisoned umbrella in 1978.

Later political shifts see the removal of many European language services, and a re-prioritization of other zones and media, notably the expansion of a TV service for Arabic in 2008 and for Persian in 2009. In 2012, the World Service leaves Bush House in the Strand, where it has been since 1941, to join other journalists in Broadcasting House.

Televising the Boat Race

Limited camera outlay in 1938 gives way to extensive contemporary coverage, with 25 cameras on land, nine on the water, and one in the air. The broadcast commands a huge international, as well as domestic, audience.

In this year

30 April
The FA Cup Final is broadcast on TV.

27 September
The European Service launches.

The Oxford and Cambridge Boat Race, first covered by radio on 2 April 1927 in an event the *Radio Times* trails as "one of the biggest treats that the BBC has provided yet", now goes televisual! The technical challenge of providing a running commentary over the length of the 6.8-km (4 miles 374 yards) course is considerable.

On radio, the first commentators – Mr Nickalls and Mr Squire – describe the race to listeners from a launch that follows the Oxford and Cambridge crews. An aerial on the launch relays the signal to two receiving stations in Barnes, from where it is sent by landline to BBC's HQ at Savoy Hill for broadcast.

For the TV coverage this year, cameras can only cover the finish line and the boathouse. Viewers have to be content with John Snagge's commentary – on radio and television – and a chart showing the progress of the race. They will have to wait until 1949 to see the whole course on television.

> "I can't see who's in the lead, but it's either Oxford or Cambridge."
>
> BBC commentator John Snagge's famous gaffe in his 1949 Boat Race commentary.

Preparing for War

"Tommy Handley was as unique in radio comedy as Charlie Chaplin was in silent film."

Comedy writer Barry Took writing Handley's entry for the Oxford Dictionary of National Biography.

Tommy Handley, at the height of his popularity in 1942.

War is coming – it's in the air… And although *It's That Man Again* (or *ITMA* as it is commonly known), launches on 12 July, before the outbreak of World War II, its title comes from a newspaper headline signalling Hitler's aggression.

However, "That Man" is also Tommy Handley, the star of the show. Handley has been a music hall and radio regular for many years, but *ITMA* makes him a household name.

The format of *ITMA*, produced by Francis Worsley and written by Ted Kavanagh, has Handley at its centre, delivering quick-fire jokes and topical references, with a supporting cast of comic characters and catchphrases. Inimitably British though it seems, it is actually designed to emulate American comedy formats. Helping Handley are popular radio favourites Jack Train, Dorothy Summers, Maurice Denham, Horace Percival, Derek Gyler and Hattie Jacques.

Among the show's classic catchphrases are: "Can I do yer now sir?"

and "Ta ta for now", from Mrs Mopp; "I don't mind if I do", from Colonel Chinstrap, who sees every enquiry as an invitation to have a drink; and "It's being so cheerful that keeps me going", from Mona Lott. The programme survives the war, but is brought to an end when Handley dies in 1949, though many of its classic catchphrases endure.

The man of the moment: Tommy Handley with Molly Weir, Linda Joyce and Diana Morrison in a scene from *ITMA*, 1946.

War Is Declared

In this year

12 July
The radio comedy *It's That Man Again (ITMA)* begins.

1 September
The BBC television service is suspended in anticipation of war.

3 September
Prime Minister Neville Chamberlain announces that Britain is now at war.

"The men – including all the musicians I knew in the band – would all be going away to fight. And I would be headed straight for the munitions factory. I was only 22. It seemed like the end of my world."

Vera Lynn, on hearing Chamberlain's announcement at her parents' home in Barking

Finally it comes: after months of tension, Britain is at war. The terrible news is broken by Prime Minister Neville Chamberlain at 11.15 a.m. on Sunday, 3 September 1939.

This is the first-ever war to be broadcast live, and everything happens much more quickly with the new rapidity of radio communications. In a five-minute broadcast on the Home Service, Chamberlain announces that as Hitler has failed to respond to British demands to leave Poland, "This country is at war with Germany." Chamberlain then goes on to say that the failure to avert war is a bitter personal blow, and that he doesn't think he could have done any more.

Other announcements quickly follow. All places of entertainment are to close with immediate effect, and people are discouraged from crowding together, unless to attend church services. Details of the air-raid warning are given and it is emphasized that tube stations are not to be used as shelters.

In London, only eight minutes later, air-raid sirens sound. Most BBC staff have already been evacuated to outside the capital. But some remain, including commentator John Snagge, who straps on his tin helmet and rushes to the roof of Broadcasting House to watch the bombs falling. It proves a false alarm, but the war has begun.

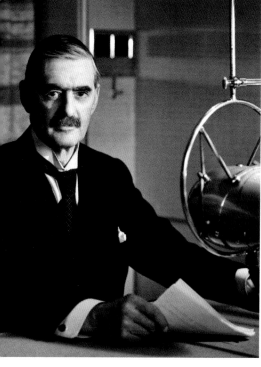

The first broadcast war: UK Prime Minister Neville Chamberlain uses BBC broadcasts to keep the nation informed.

BBC war correspondent Frank Gillard interviews Corporal R. B. Pass in France on Christmas Day, 1944.

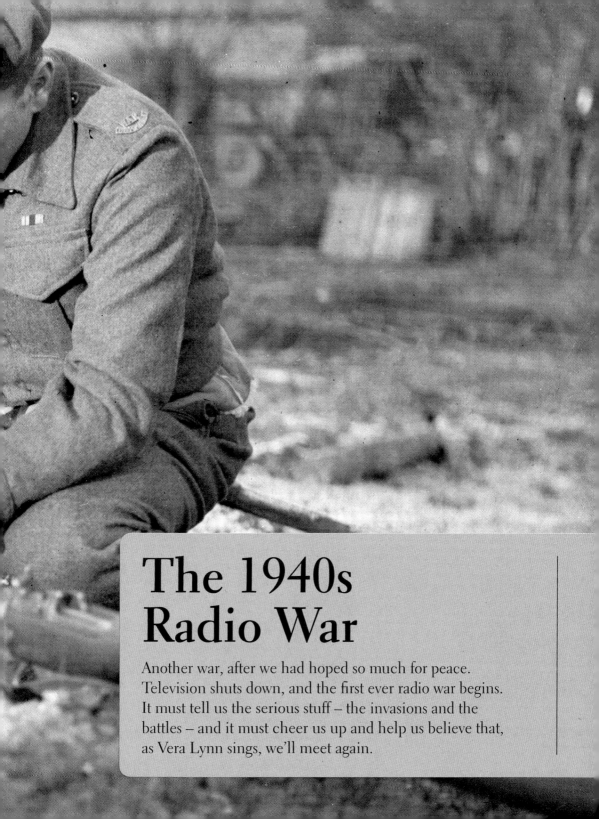

The 1940s
Radio War

Another war, after we had hoped so much for peace.
Television shuts down, and the first ever radio war begins.
It must tell us the serious stuff – the invasions and the
battles – and it must cheer us up and help us believe that,
as Vera Lynn sings, we'll meet again.

Churchill's First Broadcast Speech as Prime Minister

"He mobilized the English language, and sent it into battle."

American broadcaster Ed Murrow on Churchill as broadcaster.

Prime Minister Winston Churchill addresses the nation.

"I speak to you for the first time as Prime Minister in a solemn hour for the life of our country, of our empire, of our allies, and, above all, of the cause of freedom." So begins Winston Churchill's first radio broadcast as Prime Minister to a country at war.

Churchill is not a fan of radio but he appreciates its value in enabling him to communicate directly with the nation. And in spite of his misgivings, on 19 May he delivers some of the most stirring and memorable speeches ever given. This speech is known as "Be Ye Men of Valour" for its quotation from Book Three of the Maccabees in the King James Bible, which concludes: "Arm yourselves, and be ye men of valour, and be in readiness for the conflict; for it is better for us to perish in battle than to look upon the outrage of our nation..."

Churchill follows up this speech with others that are equally unforgettable: "Their Finest Hour" on 14 June and "The Few" – in praise of RAF fighter pilots' courage during the Battle of Britain – on 20 August.

BBC Correspondent Robin Duff and A. R. Philips of the BBC Recording Unit describe and record a dogfight over the white cliffs of Dover.

De Gaulle's Resistance Call

"La flamme de la résistance française ne doit pas s'éteindre…"
("The flame of French resistance must not be extinguished…")

The opening words of Charles de Gaulle's famous speech.

Charles de Gaulle makes his famous 'call' for Resistance fighters over the BBC airwaves.

In this year

19 May
Churchill's first radio broadcast as Prime Minister.

18 June
Brigadier General Charles de Gaulle makes his famous resistance appeal to the French nation.

23 June
Music While You Work: "Half hour's music meant specially for factory workers to listen to as they work" (*Radio Times*). Broadcast twice a day, the programme is a huge hit, gaining a memorable theme tune, "Calling All Workers" by Eric Coates in October. This morale boost for Britain's war effort will run for 27 years.

Two national leaders, both of whom become defined by their use of radio. However, unlike Winston Churchill, the French leader Charles de Gaulle takes naturally to this new communication medium, and uses it repeatedly throughout the war to mobilize the people of occupied France.

This is his first and most famous speech, which he makes from Broadcasting House on 18 June, having been given permission by the British government to speak on the BBC airwaves. It is called "L'Appel" or "The Call", and in it he summons French patriots wherever they are to join him in the fight against Nazism, to free France and reclaim the nation. Its words have a huge symbolism still for modern France: "We are all the children of 18 June," President Nicolas Sarkozy will say on the 70th anniversary of this broadcast.

Diverse Voices: Una Marson, Wilfred Pickles and Vera Lynn

"You speak good English
Little brown girl,
How is it you speak
English as though
It belonged
To you?"

"Little Brown Girl" by Una Marson, 1937

Una Marson

The first Black BBC radio producer, Una Marson joins the Corporation in March of 1941 as a Programme Assistant in the Empire Programmes department.

Originally from Jamaica, and a poet and dramatist in her own right, Marson soon begins to develop "Caribbean Voices", a weekly feature within the *Calling the West Indies* series. The segment includes poems and stories by Caribbean authors, many of whom are either unknown in the UK or only just beginning to establish a literary reputation. She gives these voices a new and international platform, and the series continues until 1958 as her legacy after her post-war return to Jamaica .

In 2009, a blue plaque is placed on the house where she lived in Camberwell, South London.

Una Marson presents
Calling the West Indies
in 1941.

Wilfred Pickles urges the nation to "Have a go!" in his popular radio show, 1947.

Wilfred Pickles

On 8 June, Yorkshireman Wilfred Pickles is the first newsreader to speak on air in a regional accent rather than using Received Pronunciation or "BBC English". This is part of a plan by Brendan Bracken, the Minister of Information, and the BBC, to avoid enemy infiltration of the media during the war by using accents that might be difficult for foreigners to understand. His closing "Goodneet" is received with mixed feelings, but the overall listener response is favourable.

Pickles will soon become a radio celebrity in his own right with his hugely popular show *Have a Go*, which runs from 1946 to 1967 and, at its peak, draws audiences of over 20 million.

Vera Lynn

She is the woman from East London who becomes the "Forces Sweetheart" of the war years. In late 1939, Vera Lynn releases her most famous song, "We'll Meet Again". Then, on 9 November, the BBC gives Lynn her own show, *Sincerely Yours, Vera Lynn*. The BBC is initially dismissive of its overt sentimentality, but the show soon gains a huge following. A year later, Lynn records the other song that will so define her singing career, "The White Cliffs of Dover".

In this year

14 January
Belgian politician Victor de Laveleye, broadcasting on BBC Europe, proposes the "V for Victory" campaign.

1 February
Una Marson becomes the BBC's first Black radio producer.

8 June
Wilfred Pickles reads the news.

9 November
Sincerely Yours, Vera Lynn

The enduring voice of Vera Lynn, 1941.

Desert Island Discs

"Dance-band leaders, actors, members of the Brains Trust, film stars, writers, child prodigies, ballet dancers and all sorts of people could be included…"

Roy Plomley's suggestions for the original guest list for *Desert Island Discs*.

Roy Plomley on his own desert island in 1982, on the 40th anniversary of the show.

In this year

29 January
Desert Island Discs

6 May
The first broadcast by the Radio Doctor, Dr Charles Hill.

On 3 November 1941, a struggling actor called Roy Plomley writes to a BBC commissioning editor: "Here is another idea for a series. If you were wrecked on a desert island, which ten gramophone records would you like to have with you – providing of course that you have a gramophone and needles with you?!"

And so begins the most popular radio show ever. The idea is speedily taken up, and Plomley himself presents the first edition on 29 January, interviewing comedian Vic Oliver in the bomb-damaged Maida Vale Studios. The success of the programme lies in its simple format, which allows for sometimes revealing interviews, alongside the musical selections.

As the show settles down, the format becomes established. Guests choose eight records to take with them, along with one luxury item – not a survival aid – and a single book. It is assumed that the Bible and the complete works of Shakespeare are already there. The castaway is then asked to pick their favourite from their eight music choices.

Written in 1930, Eric Coates' "By the Sleepy Lagoon" (with lapping waves and seagulls) introduces the show. The theme so impresses the programme's second castaway, renowned literary figure James Agate, that he makes it one of his desert island choices.

Most appearances on the island: comedian Arthur Askey (along with David Attenborough) holds the record.

Island life

Plomley presents 1,791 editions before his death in 1985. He is succeeded in the presenter's chair by Michael Parkinson, Sue Lawley, Kirsty Young and Lauren Laverne. An impressive roster of celebrity guests from all walks of life will be interviewed on the show across the years, including musicians, scientists, philosophers, politicians, actors and sports personalities. The unpredictable variety of "castaways" is one of the keys to the show's remarkable longevity.

The most popular music choice is Handel's *Messiah*. Comedian Arthur Askey and broadcaster David Attenborough will both jointly hold the record for being interviewed four times.

Baroness Floella Benjamin picks her eight discs and tells presenter Lauren Laverne about her rich and fascinating life, 2020.

Air-Raid Over Berlin

"The most beautifully horrible sight I've ever seen."

Wynford Vaughan-Thomas, recalling the vision of bombs falling on Berlin.

In this year

3 April
The first broadcast of
Saturday Night Theatre – a
run that will last 50 years.

4 September:
Wynford Vaughan-Thomas'
eyewitness account of an
air-raid over Berlin.

11 September:
Horror/ghost story drama
series *Appointment With
Fear*, introduced by Valentine
Dyall, "The Man in Black".

On 3 September, BBC war correspondent Wynford Vaughan-Thomas boards a Lancaster bomber (nicknamed "F for Freddie") with recording engineer Reg Pidsley. His mission: to capture the reality of warfare and relay it back to the listeners at home.

The team heads across the Channel, over Holland, into Germany and finally to Berlin, capital of Nazi Germany. There, Vaughan-Thomas experiences what he later recalls as "the most terrifying eight hours I've spent in my life". As he speaks, the engineer at his side is recording his emotive commentary onto a disc, then stuffing the record into his jacket, hoping to ensure the freezing temperatures inside the bomber and the jolt of the descending bombs don't break it. The relief of Vaughan-Thomas and the crew is palpable but unsaid, as they cross the coast and land safely back on British soil.

The broadcast transmitted the following day reveals the terrors and dangers of a night-time raid to the world. It remains one of the most historically important pieces of on-location journalism of the war. Vaughan-Thomas' vivid report is also used by the British government to hook in more American support for the war effort. It works.

Wynford
Vaughan-Thomas
recording
interviews in
Marseille later, in
August 1944.

1943

Wynford Vaughan-Thomas – just
one of many BBC correspondents
conveying the reality of war.

The Man Who Went to War

"It is a play about everyone who is fighting today…
all who are determined to win freedom for the world."

Paul Robeson in his stirring introduction to the play.

Actors Hall Johnson, Josh White,
Paul Robeson, Ethel Waters and
Canada Lee.

In this year

6 March
The Man Who Went to War

6 June
BBC reporters record the
events of the D-Day
Normandy landings.

This "ballad opera" for radio is created by the great Harlem
Renaissance poet Langston Hughes and BBC producer Geoffrey
Bridson. Broadcast on 6 March, it tells the story of Johnny, an African-
American soldier who decides to join the war and of his family who
stays behind. The narrative is transformed by the all-Black casting of
every role, and so the story becomes an example of a fight not just
against the Nazis but also against all oppression, including racist
attitudes in the US.

The starry cast includes Paul Robeson, Canada Lee and Ethel
Waters, with songs intercutting the action sung by Josh White.

D-Day Report

"You handful of men have been chosen to undertake the most important assignment so far known to broadcasting. Good luck!"

A. P. Ryan, Controller of BBC News

War Reporter Robin Duff in action with the new midget recorder.

The BBC War Reporting Unit at Fareham transmitter site during the D-Day period.

War is captured as it happens. The D-Day landing report of 6 June, covering the combined naval, land and air attack on Nazi-occupied France, begins with the 8.00 a.m. bulletin, as Freddy Allen reports that paratroopers are landing in France. Then, hour by hour, the BBC war correspondents relay history unfolding.

The BBC has set up a special War Reporting Unit to cover this moment: 17 reporters are trained as soldiers and embedded within the initial British and US invading forces. For correspondent Guy Byam, this will mean jumping with the 6th Airborne Division. For correspondent Howard Marshall it will involve wading ashore from a landing craft hit by a mine.

To record their "actuality" from the frontlines, the correspondents are equipped with a new piece of kit – the portable "Midget" recorder. In reality, it is far from small or light, weighing a cumbersome 18 kg (nearly 40 lb), but it does capture sound in the field as never before.

Two BBC correspondents are among the many casualties of the invasion. Kent Stevenson dies reporting on a raid over Germany two weeks after D-Day, and Guy Byam is killed in a raid over Berlin early in 1945. War Report continues nightly as the Allies push into Europe, ending after 235 editions.

VE Day

"Six long, weary and perilous years are behind us…and through them all, broadcasting was kept on the air. Today's victory is one in which everyone in the BBC can feel he or she has played a part."

Director-General William Haley in a memo to staff on VE Day

Crowds fill Piccadilly Circus in central London, as the news breaks that war is over.

On 8 May, for the first time since Coronation Day in 1937, Broadcasting House in central London is floodlit and bedecked with the flags of the 22 Allied nations. It is time at last to light up the world and celebrate.

Earlier in the day at 3 p.m., Winston Churchill has made his momentous broadcast from Downing Street, announcing the end of hostilities in Europe: "We may allow ourselves a brief period of rejoicing, but let us not forget for a moment the toils and efforts that lie ahead." A broadcast, too, from bomb-scarred Buckingham Palace by King George VI, thanking the nation. And reminding all the listeners to "remember those who will not come back".

So radio defines the beginning and the end of this war, in speech and music too – as celebratory programming takes over, and people flood the streets to dance, sing and drink.

The Light Programme

"[Children] grow more concerned from day to day about what Dick Barton may do next than about their futures or the future of England."

Letter to *The Times* in 1946, commenting on one of the Light Programme's big successes, *Dick Barton – Special Agent*.

In this year

8 May
The BBC broadcasts the VE Day celebrations to millions.

29 July
The Light Programme, specializing in music, comedy and light entertainment, launches.

1 August
Music request show for British forces overseas *Family Favourites* (later renamed *Two-Way Family Favourites*) airs, usually presented by Jean Metcalfe.

9 October
The sole programme the BBC has to make under its charter, *Today in Parliament*, is broadcast.

Replacing the General Forces Programme, which ran during the war, the new Light Programme launches on 29 July, offering "all that is best in radio entertainment from nine-o'clock in the morning till midnight", in the words of its Chief Assistant Tom Chalmers. The network quickly becomes the most listened to BBC radio service.

The first day is typical of the station's varied output. It begins with *Transatlantic Quiz*; a short story by W.W. Jacobs; *Variety Bandbox*, a mix of comedy and music; and concludes with soothing music shows *In a Sentimental Mood* and *Songs of Three Decades*.

Over the years, the Light Programme introduces many hugely popular shows, including *Family Favourites* (1945), *Women's Hour* (1946), *Dick Barton – Special Agent* (1946), *Mrs Dale's Diary* (1948), *Hancock's Half Hour* (1954) and *Round the Horne* (1965).

In 1967, the Light Programme is renamed Radio 2, continuing the same highly popular mix of music and entertainment and becoming the most listened-to radio network in Europe.

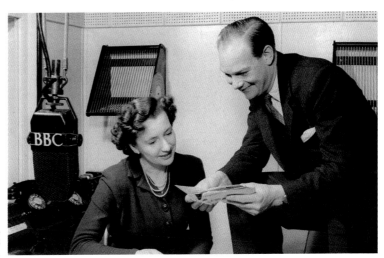

Jean Metcalfe and Cliff Michelmore, presenters of *Family Favourites* in 1949; the couple will marry in 1950.

For the Children

"Here comes Muffin, Muffin the mule,
Dear old Muffin, playing the fool,
Here comes Muffin, everybody sing:
Here comes Muffin the mule."

Theme tune to *Muffin the Mule* by Annette Mills.

Television is back! It was switched off during the war years, but on 9 June it returns with an ambition to fill its schedule with exciting new programmes for audiences of all ages, including the very young.

The opening show of *For the Children*, features The Hogarth Puppet Circus (who will produce many of the BBC's early Children's TV feature puppets). Then in October of this year, they return with their first star performer: the clip-clopping Muffin the Mule, accompanied by presenter Annette Mills, sister of the actor John Mills.

In 1950, the BBC will create a specific Children's TV department, and, jump forward 50 years, in 2002 it will launch two dedicated digital channels, CBBC and CBeebies, creating a plethora of popular shows for young audiences and their families.

Annette Mills with Muffin the Mule, 1952.

Woman's Hour

"When we did a programme on the change of life... we had hundreds of letters from ladies who thought they were suffering alone."

Olive Shapley, presenter of *Woman's Hour* in the 1940s, speaking about the programme's coverage of the menopause.

Emma Barnett, presenter of *Woman's Hour* from 2021.

This flagship series for women begins its long and continuous run on 7 October. The BBC has made programmes for women listeners before, including a rich strand of "talks" programmes in the 1920s and 30s. However, *Woman's Hour* establishes a firm female voice in the schedule, and, through its longevity, charts the changing role of women in Britain. The programme tackles current affairs and contemporary topics, as well as some hitherto "taboo" subjects such as the menopause, divorce and illegitimacy.

Bizarrely for us now, the first edition is presented by a man: Alan Ivimey, who according to *Radio Times*, is a specialist "in writing for and talking to women". The presenter soon changes to the more appropriate Mary Hill, to be followed by a succession of popular and female presenters, such as Jean Metcalfe, Sue MacGregor, Jenni Murray and Jane Garvey. *Woman's Hour* will remain a cornerstone of BBC Radio's weekday schedule for decades to come, providing a female perspective on the world via a distinctive mix of challenging and lighthearted content.

T. Holland Bennett interviews actress Deborah Kerr and Elsie May Crump and Margaret Bondfield.

The Last Night of the Proms

"The Last Night of the Proms is not just another concert. It's an event all about its audience: its joyous, inflatable-toting, flag-waving, singing-and-knee-bobbing crowd of prommers, who whistle, whoop and drink their way through the evening."

Guardian review, 2020

Flag waving and singing from the Last Night Promenaders, 1999.

In this year

13 September
The Last Night of the Proms

2 November
Round Britain Quiz

The Proms receives its last night finale for the first time on 13 September. This comes about through a combination of two events: the arrival of TV cameras for the first time, and the appointment of the flamboyant Malcolm Sargent as lead conductor. Sargent – like his original predecessor Henry Wood – is a keen ambassador of classical music for all, and he wants to get the Proms audience up and singing.

It is Sargent who insists on the fixed and populist pattern for the Last Night repertoire, ending with Sir Henry Wood's *Fantasia of British Sea Songs*, concluding with "Rule, Britannia!"; "Land of Hope and Glory", sung to the theme from Elgar's *Pomp and Circumstance March No. 1*; and lastly William Blake's "Jerusalem" with music by Sir

68

Hubert Parry, belted out by the promenaders with much flag-waving for the cameras.

Sargent actively encourages the boisterous exuberance of the crowd, insisting that "if people get as enthusiastic about music as they do about football, it is all to the good". The tradition becomes a part of the calendar celebration of Britishness, though some critics see the broadcast as jingoistic.

In 2020, controversy arises over the planned removal of the words of "Rule, Britannia!" because of their negative connection with the former British Empire. However, in the end, the words remain, sung by the BBC singers.

The flamboyant Malcolm Sargent in action in 1947.

Round Britain Quiz

"If you know the answer and say it straight out, it's no fun. You have to amble towards it, and the fun always comes when you get lost on the way."

Regular panellist Irene Thomas.

Round Britain Quiz is the successor to *Transatlantic Quiz* (presented by one Alistair Cooke) and remains the longest-running radio quiz in the world.

First broadcast on 2 November, the format is unchanging: a two-person London team takes on a two-person regional team, attempting to answer cryptically worded clues. Famous quiz masters include Lionel Hale and Gilbert Harding. Radio 4 currently describes it as "Radio's most fiendish quiz!"

Quizmaster Gilbert Harding questions on-air contestants, while producer Alun Davies looks on, 1949.

The First Televised Olympic Games

"Under the blazing summer sky and in the presence of their majesties the King and Queen, and in a spirit of good fellowship that stirred every one of the 80,000 spectators..."

Opening commentary for the BBC film about the coverage of the 1948 Olympic Games

In this year

5 January
The first *Television Newsreel* airs.

5 January
The first episode of radio serial *Mrs Dale's Diary*.

23 March
Radio comedy show *Take It From Here*.

1 May
Top of the Form, a radio quiz for secondary-school children.

29 July
The Olympic Games are televised.

12 October:
Radio discussion programme *Any Questions*, chaired (for 19 years) by Freddie Grisewood.

Sport is always a big draw for outside broadcasts, and on 29 July the BBC achieves the coup of being the first-ever broadcaster to televise the 1948 Olympic Games directly into people's homes.

The Olympics are held in Wembley, and so the BBC sets up a bespoke production centre nearby, providing facilities there for the broadcasters of all 61 competing nations, so they can be seen and heard around the world. After nearly two years of planning and installation, the BBC eventually screens 68 hours of coverage (as well as providing radio commentary), using the most advanced cameras available.

The number of people with TV sets is still small – only about 100,000 households, and mainly in London – but the promotional impact of the Olympics is significant, with athletes such as track and field star Fanny Blankers-Koen and long-distance runner Emil Zátopek capturing the public imagination.

The Olympics are ultimately a triumph for the BBC Outside Broadcasts Department, who will use the expertise gained when they embark on their next challenge: the Coronation of Queen Elizabeth II in 1953. The Olympic Games will return to London in 2012, when the BBC launches the first-ever completely digital Olympics and coverage reaches a record 51 million viewers in the UK, and 4.8 billion across the globe.

Jack Crump, Pat Landsberg and Ian Orr-Ewing inside the commentators' box at the Empire Stadium, Wembley.

ITALY   IRAQ   IRAN   INDIA   ICELAND   HUNGARY   FRANCE   FINLAND

A BBC camera shows the march past of the BBC team during the opening ceremony of the Olympic Games, 1948.

The First Television Weather Forecast

"Earlier today, apparently, a woman rang the BBC, and said she'd heard there was a hurricane on the way. Well, if you're watching, don't worry, there isn't."

Michael Fish's ill-fated words of 15 October, 1987, a few hours before the Great Storm breaks.

In this year

31 January
The Three Hostages by John Buchan is the first-ever *Book at Bedtime*.

6 March
The Billy Cotton Band Show, heralded by Billy's "Wakey, wakey!" catchphrase.

29 July
The first TV weather forecast.

29 September
Peter Dimmock introduces ballroom-dancing competition *Come Dancing*.

On 29 July of this year, weather comes to television, but not as we will come to know it in the future. There is no presenter, merely a rather crudely visualized weather map and an invisible voice.

Weather broadcasts have been banned during the war years, for fear that they will supply the enemy with vital information for air attacks. The ban is lifted in October 1944, and a BBC News bulletin reports ironically: "Most people will have cause to remember it because in most parts of the country it just rained and rained…"

"A good day for drying"

On 11 January, 1954, George Cowling will become the first weather forecaster in vision. Cowling's style is a long way from the informal style of current weather presenters, but he adds a personal touch when he says, "Tomorrow will be a friendly wash day and a good day for drying." Cowling uses charcoal sticks to draw weather features on two charts, one for today's and one for tomorrow's forecast.

The first weather forecaster in vision – George Cowling in 1954.

Weather forecaster Michael Fish in 1987.

All weather forecasters are professional meteorologists, employed by the Meteorological Office (Met Office). It is not long, however, before weather forecasters become television personalities in their own right. Michael Fish will earn his place in television history when he tells the nation that it's not a day for hurricanes, just as the 1987 Great Storm breaks...

In 1974, Barbara Edwards becomes the first female forecaster on television. Years later, in 2015, one of the nation's most popular forecasters, Carol Kirkwood will perform on *Strictly Come Dancing*, and, in 2021, North West Tonight's weatherman Owain Wyn Evans will break *Children in Need*'s 24-hour challenge records when he raises £3 million drumming for the charity. Weather forecasting has come a long way from a static map and an anonymous voiceover.

Weather presenter Carol Kirkwood takes to the *Strictly Come Dancing* stage in 2015.

The State Coach returns to Buckingham Palace, following the Coronation of Her Majesty Queen Elizabeth II, on 2 June 1953.

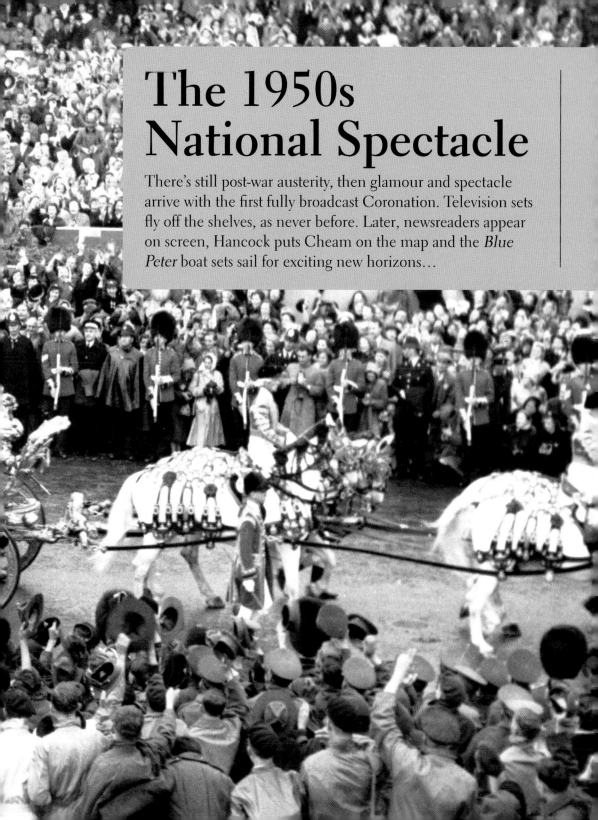

The 1950s
National Spectacle

There's still post-war austerity, then glamour and spectacle arrive with the first fully broadcast Coronation. Television sets fly off the shelves, as never before. Later, newsreaders appear on screen, Hancock puts Cheam on the map and the *Blue Peter* boat sets sail for exciting new horizons…

General Election Results Televised

"An important landmark in the history of vision broadcasting as a medium of news information."

Radio Times edtiorial

In this year

16 January
Listen With Mother airs.

23 February
The General Election results are broadcast on TV.

June
Radio comedy *Educating Archie* debuts, starring ventriloquist Peter Brough and his doll Archie, as well as a host of future comedy stars, including Tony Hancock, Harry Secombe, and Hattie Jacques.

11 July
Puppet character Andy Pandy debuts on the TV show *For the Children* (later *Watch With Mother*).

27 August
The first TV broadcasts from continental Europe.

26 October
Experimental broadcast from the House of Commons.

Because of the longstanding sensitivity around politics and broadcasting, it has taken almost 30 years for active democracy to have a voice on TV or radio, and even then there is no coverage of the campaign leading up to the results.

But finally, on 23 February 1950 the first General Election results are screened on an extended television service. "Elections results as they come in!" shouts the *Radio Times*.

The programme has specialist commentary from R. B. McCallum, David Butler and Chester Wilmot, and the final results are heralded by "one of the most important studio devices…best described as a cricket-type scoreboard, and this will show the progress of the parties". Added interest comes from an outside-broadcast truck parked in Trafalgar Square, depicting the crowds gathered to watch the latest returns projected on a huge screen.

And the final result? Labour under Prime Minister Clement Atlee remains in power, but with a slender majority of just five MPs.

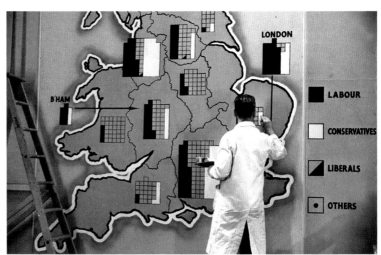

General Election results are brought to life via new studio graphics.

TV From the House of Commons

A brief experiment to capture live debate from the House of Commons occurs on 26 October, but it is not publicly visible. It will take some decades for this to happen, as there are concerns that television will turn the serious matter of politics into a sideshow.

Live radio broadcasts from the Commons will be trialled as a month-long experiment in 1975. Television comes to the House of Lords in 1983, and finally, in 1989, it reaches the House of Commons.

The BBC Home News department manages a feed from the House of Commons.

Experienced BBC correspondent Wynford Vaughan-Thomas interviews direct from the House of Commons.

The Archers

"Whatever else is going on in the world, or in our lives,
there'll always be the single wicket competition, Lynda's Christmas show,
harvest and the flower and produce show in Ambridge. And somebody
somewhere will be putting the kettle on."

I Love Archers, Radio 4 site.

Patricia Greene as Jill Archer,
the matriarch of the family. She
begins playing the role in 1957, and
will become the longest-serving actor
in a soap opera in any medium.

The first episode of *The Archers*, the longest-running daily serial in the
world, is broadcast on 1 January 1951. It actually started a year earlier
as a pilot programme on the Midlands Home Service, directly
targeting the farming community. However, it quickly becomes clear
that there is a far wider audience for this "everyday story of country
folk", as the *Radio Times* describes it.

The original idea for the serial comes out of post-World War II
agricultural depression. At a meeting of farmers in Birmingham to
discuss how best to motivate innovation in farming, one audience

A family gathering
of the Archers in 1958.

member pipes up: "What we really want is a farming Dick Barton!"
And so the writers of *Dick Barton*, a popular secret-agent radio drama,
are brought in to initiate the scripting process, under the watchful eye
of producer Godfrey Basley.

Forever Ambridge

Set in the fictional village of Ambridge in the county of Borsetshire,
the show grows a huge fanbase over the years, summoned to the radio
by its instantly recognizable theme tune, "Barwick Green" by Arthur
Wood. When commercial competition arrives from ITV in 1955, The
Archers stages the dramatic death of leading character Grace Archer
to distract the audience.

The serial's fans are loyal to its cosy atmosphere of everyday village
life epitomized by the pronouncements of old Walter Gabriel (Chris
Gittins). However, in later years they will be confronted with the
realities of foot and mouth disease, mental health conditions, domestic
abuse and even murder.

Attracting an audience of five million listeners per week, *The
Archers* becomes the most popular non-news show on Radio 4.

The ever-expanding Archers
community in 2006.

Crazy People (The Goon Show)

"I was 12 when *The Goon Show* first hit me, 16 when they finished with me. Their humour was the only proof that the world was insane..."

Former Beatle John Lennon, *New York Times*, 30 September 1973.

The inimitable Goons, on the day of the broadcast of May 27 1951, and no smoking being allowed in the studio *(left to right)*: Harry Secombe is attempting to swallow a cigarette; Michael Bentine is still growing his busby; Spike Milligan, not owning a tie, has turned his back to the audience; and Peter Sellers is about to telephone "The Light of His Life".

Starring Spike Milligan, Peter Sellers, Harry Secombe and Michael Bentine, *Crazy People* hits the airwaves with goonish flair on 28 May 1951. These four men have all lived through World War II, and they respond with a unique anarchic humour to the changed world they return to.

The first episode includes many of the comedy's typical features, from ridiculous sound effects and surreal situations to a pell-mell of catchphrases and hyperbolic characters with extravagant names, such as Sir Harold Porridge and Ernie Splutmuscle. To be followed later by the idiotic Eccles, squeaky-voiced Bluebottle and gullible Neddie Seagoon, respectively voiced by Spike Milligan, Peter Sellers and Harry Secombe.

Spike Milligan is the main writer and creative force, but the show makes solo stars of all its players, who, post 1960 when the show finishes, carve out major TV and film careers. The impact of *The Goon Show* will be huge, influencing the likes of *Monty Python's Flying Circus* (1969), *The Hitchhiker's Guide to the Galaxy* (1981), and *The League of Gentlemen* (1999).

80

What's My Line?

In this year

1 January
The Archers

28 May
Crazy People

18 June
TV's first visit to
The Royal Tournament,
Earl's Court, London.

16 July
What's My Line?

"Mystery Guest, will you enter and sign in, please?"

The opening words of *What's My Line?*

This hugely popular panel game turns on the challenge of guessing a contestant's profession, based on a short mime of their job. During the show, a celebrity guest appears and the blindfolded panel have to identify them, too, often penetrating absurd accents the guest has adopted to disguise their voice.

Beginning on 16 July, *What's My Line* will run for 12 years until 1963, with Eamonn Andrews at the helm. It returns again in 1973 with David Jacobs as host, and later on ITV with Eamonn Andrews again.

A guessing game for Marghanita Laski, Jerry Desmonde,
Elisabeth Allen and Gilbert Harding, 1952.

Programmes for Children

"Flobadob, ickle Weed."

Bill and Ben say hello to Little Weed in Oddle Poddle, the language invented for the show.

Bill and Ben come out to play watched over by Little Weed.

In this Year

14 April
The Vision Electronic Recording Apparatus (VERA) videotape recorder is developed.

15 December
The Flowerpot Men

Enter Bill and Ben, the Flowerpot Men. They live in a shed at the bottom of the garden, together with Little Weed ("We-e-e-e-d") who lets the duo know when it is safe to come out and play, and when the gardener is returning and they must vanish back into their pots. Bill and Ben are famous for their nonsensical Oddle Poddle language, invented by Peter Hawkins, who will later supply the voices for *Doctor Who*'s Daleks!

Debuting on 15 December, Bill and Ben join the earlier success of Andy Pandy, who first arrived in his trademark striped blue and white costume and floppy hat in 1950. Also targeted at the under fives, and initially an experimental programme written by school teacher Maria Bird – viewers were invited to send in their comments – *Andy Pandy* proves such a success that the show is developed into 26 episodes.

Andy Pandy and Teddy say goodbye at the end of the show.

It will continue on television until 1969, and so is remembered by generations of children. It is only removed then because the black and white print finally wears out.

The Flowerpot Men and *Andy Pandy* are separate programmes that will eventually combine together with *Rag, Tag and Bobtail*, *Picture Book* and *The Woodentops* under the umbrella title of *Watch With Mother* (1955). They create – for the first time ever – a regular programming slot (1.45 p.m.), every day of the working week, for the very young at home.

Are you sitting comfortably?

Watch With Mother is the TV complement to *Listen With Mother*, launched on the radio on 16 January 1950. Again targeting the very young around the post-lunchtime slot, it provides a mix of nursery rhymes, stories and music (the programme's wistful piano theme is "Berceuse" from Gabriel Fauré's *Dolly Suite*). *Listen With Mother*'s presenters – including Daphne Oxenford, Julia Lang and Dorothy Smith – use the intimate quality of radio to appeal directly to children; an ad lib by Julia Lang launches one of the most famous BBC catchphrases ever: "Are you sitting comfortably? Then I'll begin."

Julia Lang, accidental creator of one of the most famous of all BBC catchphrases.

The power of television to captivate: the first generation of viewers is spellbound by *Andy Pandy*.

The Coronation of Queen Elizabeth II

"My biggest thrill was being awakened one morning by a fanfare of trumpets. I couldn't make out what it was at first. Then I realized we were hearing the Coronation broadcasts from the Abbey."

A petty officer on board the British submarine HMS *Andrew* in the depths of the Atlantic Ocean.

For the very first time, the crowning moment is seen on television.

Behind the scenes

"There was a rule that no camera could be closer than 30 feet to the Queen, so we did a test in the Abbey and I put in a two-inch lens... and the Queen looked miles away. Of course, when the Coronation came, I knew that I was going to use a 12-inch lens, which would give the best close-up of the Queen you'd ever seen!"

Peter Dimmock explains how he persuaded Establishment figures to let cameras into Westminster Abbey.

The Coronation of Queen Elizabeth II, broadcast live on 2 June 1953, is the nation's first big moment of post-war glamour, after years of privation and hardship. It has a young and photogenic queen as its central player, a parade of British pomp and circumstance surrounding it, and underpinning it all, a sense of hopeful new beginnings to lift the spirit.

The Coronation also transforms broadcasting. Over 20 million people watch the Coronation on television, outnumbering the radio audience for the first time. Following the public's enthusiastic reaction to the limited broadcast of George VI's Coronation in 1937, the BBC knows the event will be extremely popular. But it cannot imagine that the 1953 Coronation will mark television's coming of age as a mainstream medium – as well as the modernization of the monarchy itself.

"The year that made the day"

The Coronation is officially announced on 7 June 1952, giving the BBC less than a year to plan by far the biggest and most technically challenging outside broadcast in its history. In addition, there is the extra responsibility of providing international coverage to the Commonwealth and wider world.

The project does not begin well. There is resistance from the Establishment to cameras covering the ceremony inside Westminster Abbey, in particular the sovereign's anointing. However, after lengthy negotiation and not a little sleight of hand from the BBC's Outside Broadcasts supremo Peter Dimmock, plus the goodwill of the Queen herself, planning proceeds apace.

Just three cameras televised the 1937 Coronation. By 1953, 21 cameras are placed at the five key sites: Victoria Memorial (covering

Buckingham Palace), Hyde Park, Victoria Embankment, the Colonial Office, and Westminster Abbey, where a temporary control room handles five cameras.

Commentary from key points along the procession route begins with a prearranged verbal signal – "The coach moves!" – from popular *Forces Favourites* announcer Jean Metcalfe. Over 100 commentators describe the scene as the royal coach passed by, while the key events of the ceremony in Westminster Abbey are immortalized by the BBC's leading political commentator Richard Dimbleby.

Bernard Braden and Brian Johnston, just two of the many Coronation commentators.

Twenty million watch…

News of the Coronation broadcast boosts sales of televisions to an estimated 2.5 million. Nevertheless, few people own a set in 1953 and people crowd into the living rooms of friends who have one. Sets are also rigged up in church halls and other public places.

The BBC's coverage is relayed around the world, hugely enhancing the Corporation's international profile. Recordings are despatched by helicopter and jet to Canada, enabling audiences in North America to watch proceedings on the very same day as their British counterparts.

Broadcasting House is lit up to mark the occasion.

Multiple cameras capture the progress of the State Coach, on its route to Westminster Abbey.

The Quatermass Experiment

"A landmark of science fiction and the cornerstone of the genre on British television."

BFI Screenonline

Prof. Bernard Quatermass (Reginald Tate, far left) surveys distressed astronaut Victor Carroon (Duncan Lamont), little realizing the horrors to come.

Britain's first sci-fi serial, *The Quatermass Experiment*, literally explodes onto our TV screens on 18 July 1953. Written by Nigel Kneale and produced by Rudolph Cartier, it tells the story of the first crewed rocket launch carried out by Professor Quatermass (Reginald Tate). Crash-landing, the rocket's only survivor becomes contaminated with an alien life-form capable of infinite reproduction. After a tense manhunt, the alien being is cornered by Quatermass in Westminster Abbey and destroyed just in time to save the world from disaster.

Hugely successful, *The Quatermass Experiment* spawns four TV series, a radio show and three Hammer horror films.

Panorama

The *Panorama* logo, 1996.

"We hope to have as an opening a moving film shot of London becoming smaller and smaller until it is virtually a panorama..." This is first editor Dennis Barden's early description of what was to become the world's longest-running current affairs programme. It highlights his vision of *Panorama* as a programme that will add topicality, context and significance to its treatment of contemporary news.

However, *Panorama* begins badly, and its opening show, broadcast on 11 November 1953, is very nearly its last, owing to technical

Trusted *Panorama* anchor Richard Dimbleby in 1958.

In this year

16 February
First showing of the "Potter's Wheel Interlude", one of several film sequences used to fill intervals between programmes or breakdowns, common at the time.

16 April
Chancellor of the Exchequer Rab Butler delivers the first televised budget broadcast.

2 June
The Coronation of Queen Elizabeth II is televised.

18 July
The Quatermass Experiment

20 July
Old-time music hall series *The Good Old Days*, in which the audience is as much a part of the show as the performers, airs.

21 September
Radio sci-fi series *Journey Into Space*.

11 November
Panorama

hitches and a shaky performance by host Patrick Murphy. The programme is taken off air for a month; its second edition is fronted by Max Robertson.

In 1955 it is revamped under the penetrating gaze of Grace Wyndham Goldie – who will become Head of BBC News and Current Affairs in the 1960s. Goldie refocusses the programme's output, extends its running time to an hour, and casts Richard Dimbleby as its anchor, along with such celebrated reporters as Robert Kee, James Mossman, Robin Day and Ludovic Kennedy.

TV "firsts"

Panorama will soon rack up a number of television firsts, including broadcasting the birth of a baby in 1957, interviewing the Duke of Edinburgh in 1961, and, not forgetting the most famous TV spoof ever – the April fool's spaghetti crop film in 1957.

One of its most celebrated, and notorious, coups is the interview with Princess Diana in 1985 conducted by Martin Bashir, where she reveals the truth of her flawed marriage to Prince Charles. Later, in 2021, this will be the subject of an independent investigation and a subsequent apology from the BBC to her sons, Princes William and Harry.

Panorama will go on to win countless BAFTA and Emmy awards for its in-depth journalism, defining its mission as "revealing the truth about the stories that matter".

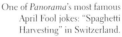

"The best programme… on television since the Coronation."

Viewer comment on *Panorama*'s coverage of an atomic bomb test.

One of *Panorama*'s most famous April Fool jokes: "Spaghetti Harvesting" in Switzerland.

Under Milk Wood

"Time passes. Listen. Time passes. Come closer now."

First Voice from *Under Milk Wood* by Dylan Thomas

Poet Dylan Thomas in front of the BBC microphone for an earlier broadcast, 1948.

From its arresting opening words ("To begin at the beginning…") spoken by the young Richard Burton, *Under Milk Wood* is one of the most astonishing pieces of radio drama ever written. Created by the poet Dylan Thomas, it reveals the hidden lives of a small Welsh village Llareggub ("bugger all" backwards!) in language that is fresh, exciting and revelatory.

Thomas has actually been working on-and-off on the play for nearly 20 years. It was premiered in the USA a year earlier, and Thomas wrote to his wife Caitlin with palpable relief: "I've finished that infernally eternally unfinished 'Play', and have done it in New York with actors."

The drama is populated by some of the best-loved characters in literature, from blind Captain Cat to Polly Garter, Reverend Eli Jenkins to No Good Boyo. Lyrically written, it's both riotously funny and deeply moving, and although firmly rooted in place, the universality of the characters shines through – which is why it's never been out of print, has been translated into over 30 languages and is regularly performed all over the world.

Sadly, the author never gets to hear the BBC Radio premiere on 1 January, as he dies of alcohol poisoning in New York on 9 November 1953, aged just 39.

Richard Burton in the studio with producer Douglas Cleverden for a later recording of *Under Milk Wood* in 1963.

Nineteen Eighty-Four

George Orwell worked at the BBC from 1941–43, and elements of its corporate bureaucracy are reflected in his novel *Nineteen Eighty-Four*.

George Orwell's dystopian novel *Nineteen Eighty-Four* arrives with a clamour on British TV, starring Peter Cushing as its Everyman hero Winston Smith and Yvonne Mitchell as his lover Julia. Apart from some location scenes, it is performed live.

Published in 1949, and adapted for the small screen by Nigel Kneale, its vision of an authoritarian state shocks the nation, especially its grim scenes of torture in the notorious Room 101. There are complaints from Parliament and in the press, but these only stoke interest in the drama, so much so that its repeat slot a few days after the first transmission draws the largest viewing audience since the 1953 Coronation. *Nineteen Eighty-Four* will be remade by the BBC in 1965, with no controversy this time.

Winston Smith (Peter Cushing) begins an ill-advised affair with Julia (Yvonne Mitchell).

Sports Review of the Year

"I thank all concerned for the award. I had a good time boxing. I enjoyed it and… I may come back!."

Muhammad Ali, accepting his award while battling Parkinson's disease.

Peter Dimmock
(1920–2015)

Peter Dimmock is one of the pre-eminent figures of BBC outside broadcasts in the 1940s and 50s. He covers the 1948 Olympic Games and the 1949 Oxford and Cambridge Boat Race, and masterminds coverage of the Queen's Coronation, becoming Head of Outside Broadcasts in 1954. He is also a famous face on air, fronting *Sportsview*, *Grandstand* and *Sports Review of the Year*. He subsequently works for the the BBC's commercial wing and the American Broadcasting Company.

Sports broadcasting really takes off on television in the 1950s, under the inspired leadership of Peter Dimmock, the Outside Broadcast mastermind of the 1953 Coronation, and Paul Fox – later to become Managing Director of BBC Television.

Earlier in 1954, they launch *Sportsview*: "Welcome sports fans to this first edition of your own programme," says Dimmock. Then in December, they round the year up with what will become one of the BBC highlights of the sporting calendar, *Sports Review of the Year*, which changes its name in 1999 to the iconic *Sports Personality of the Year*.

Dimmock presents the first award on air to Christopher Chataway, who wins it for his last-gasp, world-record 5,000-metre victory over Russian runner Vladimir Kuts in a London vs Moscow match on 13 October. The presentation is watched by over 12 million viewers. Roger Bannister is the runner-up after becoming the first man to run a mile in under four minutes.

Other winners over the years will feature a roll call of British sporting talent, in every arena. The award for Sports Personality of the Century is won in 1999 by the boxer Muhammad Ali.

The coveted Sports Personality of the Year trophy.

Sports Personalities of the Year (*clockwise from top*): The first award winner, athlete Christopher Chataway, 1954; skating duo Jayne Torvill and Christopher Dean, 1984; middle-distance athlete Kelly Holmes, 2004; long-distance runner Mo Farah, 2017; tennis superstar Andy Murray, 2015; heavyweight champions Lennox Lewis and Muhammad Ali, The Sports Personality of the Century, 1999.

In this year

25 January
Under Milk Wood

2 November
Hancock's Half Hour

12 December
Nineteen Eighty-Four

21 December
Young zoologist David Attenborough makes his first BBC appearance in *Zoo Quest*, a natural history series that sees him questing for exotic animals overseas to bring back to London Zoo.

30 December
The first *BBC Sports Review of the Year* award ceremony.

Dixon of Dock Green

"Evenin' all."

P. C. Dixon's regular salutation to viewers at the start of each show.

The reassuring salute of Constable George Dixon (Jack Warner), marking the end of every programme.

Police Constable George Dixon (played by well-known film actor Jack Warner) is the comforting face of law and order in 1955. First airing on 9 July, Dixon patrols his beat of Dock Green, located somewhere in East London, and issues calming homilies direct to camera at the end of each programme.

His character was originally created for the 1949 Ealing film *The Blue Lamp*, co-starring a young Dirk Bogarde as a trigger-happy criminal. Dixon dies in the film, but resurfaces here in a role that will define him for over 20 years and 432 episodes. In fact, he is so closely identified with Dixon that when Jack Warner himself dies, officers from the Metropolitan Police carry his coffin. Written by Ted Willis, the show also features the lilting harmonica of Tommy Reilly playing its nostalgic theme tune "An Ordinary Copper".

Much edgier police dramas are on the horizon – notably *Z Cars* (1962), set in Liverpool, and *Softly, Softly* (1969) set in Bristol, with scenes shot on location. Both series will feature much racier and more realistic storylines; but *Dixon of Dock Green* persists as a symbol of a safe post-war Britain, or rather of the post-war Britain that we would like to believe in.

This Is Your Life

"The greatest idea to have come out of television on either side of the Atlantic."

Presenter Eamonn Andrews on *This Is Your Life*.

Eamonn Andrews holds the famous red book.

Based on a highly successful American TV show, *This Is Your Life* crosses the Atlantic to BBC screens on 29 July this year, and becomes an immediate hit, running monthly, fortnightly, then eventually weekly by its third series. The show runs for nine years, before transferring to ITV, then returning to the BBC in 1994. Its simple but revelatory format is the secret of its success, as the presenter with the big red book doorsteps an unknowing personality to take them through the significant moments and people of their life. Popular presenter Eamonn Andrews is the show's first host, and he is also – by accident rather than design – *This Is Your Life*'s first subject!

In 1961, footballer Danny Blanchflower becomes the only celebrity to flatly refuse to take part when surprised by the host.

Crackerjack!

"It's Friday... It's five to five... It's Crackerjack!"

The famous opening to this much-loved children's TV show.

The *Crackerjack!* team in 1969 (*from left*): Rod McLennan, Jillian Comber, Michael Aspel, Frances Barlow and Peter Glaze.

This is the children's show that loves to get its live audience up and active, right from the word go! Whether it's shouting "Crackerjack!" every time the name of the show is mentioned, singing along with pop-star guests, or competing in fun games – all for the chance of winning a famous *Crackerjack!* pencil.

Commencing on 24 September, the show will be presented by a host of popular presenters over the years, including Leslie Crowther, Eamonn Andrews and Michael Aspel. It survives until 1984, before making a comeback in 2020, presented by Sam and Mark.

BBC Newsreaders in Vision

"We were not to be seen reading the news… Any old still, map or chart was preferable to a human face."

Richard Baker on the previous non-visibility of the BBC newsreader.

News gains a new profile, and for the first time, on 4 September, Richard Baker and Kenneth Kendall – to be joined shortly afterwards by Robert Dougall – become the face of BBC News.

News has been growing in screen visibility from the late 1940s onwards. On 5 January 1948, the first Television Newsreel launched, with its iconic opening titles featuring the Alexandra Palace transmission mast. Adapted from the already hugely popular cinema newsreel format, it covered major news stories as well as sport, cultural and quirky human interest stories. Initially televised bi-weekly, this was increased to three editions in December 1950 and five editions in June 1952.

Then in July 1954, Richard Baker's voice is heard off camera behind a filmed view of Nelson's Column: "Here is an illustrated summary of the news. It will be followed by the latest film of happenings at home and abroad". A couple of months later, he appears in vision, and our relationship with news changes forever.

Why has it taken so long? Dating back to the earliest days of the BBC, there was a tradition that important news broadcasts should not be distracted by the personality of the announcer; it takes many years to overturn this rule. The other trigger is undoubtedly the arrival of the BBC's commercial rival, ITN in 1955, with its strong team of newsreaders as well as an engaging entertainment offer. Broadcast monopoly ends for the BBC, and with it the realization that the Corporation needs to up its game to keep audiences watching.

Richard Baker reads the TV evening news (off camera) on 5th July 1954.

Newsreader Robert Dougall clearly in vision in 1966.

The BBC's First Female Newsreader

Jouranlist Nan Winton – unfairly rejected
as a newsreader by the public.

"In Italy and Spain, they have women
newsreaders...
We're afraid of that here."

Nan Winton's comments following her dismissal.

On 20 June 1960, Nan Winton will become the first woman to read the news on BBC television, following in the footsteps of Barbara Mandell on ITV. Despite being an experienced broadcaster who has previously worked on *Panorama* and other high-profile news programmes, Winton does not last long in her role. Audience research at the time finds that a significant number of viewers do not accept bulletins read by a woman, as they deem women are "too frivolous to be the bearers of grave news". Nan Winton leaves, understandably furious.

Hancock's Half Hour

> "It's both funny and sad, which seem to me the two basic
> ingredients of good comedy."
>
> Tony Hancock on in-depth interview show *Face to Face*, 1960

**Ray Galton (1930–
2018) and Alan
Simpson (1929–2017)**

Ray Galton and Alan
Simpson are one of the most
successful comedy writing
teams of the BBC. In 1948,
the two meet as teenagers
recovering from TB
in a medical facility, and start
work as comedy gag writers.
Their two greatest triumphs
will be *Hancock's Half Hour*
(1954) and *Steptoe and Son*
(1962), both distinguished by
a tragi-comic mix of thwarted
ambition and aspirational
delusion.

Tony Hancock translates his comic genius to television on 6 July,
creating a new tradition of radio to TV transfer. The Hancock world is
located at 23 Railway Cuttings, East Cheam – anonymous, post-war
suburbia – where Anthony Aloysius St John Hancock, to give him his
full fictional name, aspires to greater recognition.

Television suits Hancock's repertoire of lugubrious facial
expressions, and the programme quickly becomes the most-watched
show on TV. Scripted by Ray Galton and Alan Simpson, it is
performed live (until 1959). Co-star Sid James (of *Carry On* films
fame), helps Hancock master the subtler
mysteries of TV acting. Occasionally
Hancock struggles to learn his lines
and resorts to taping them to
props on set.

Hancock's Half Hour runs for
57 episodes until May 1960
– with the radio show
running in tandem on
the Light Programme.
The final BBC TV series,
Hancock, launches in
1961, and includes
such masterful
sketches as "The
Radio Ham" and
"The Blood Donor".

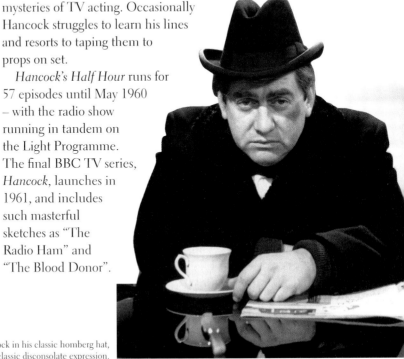

Hancock in his classic homberg hat,
with his classic disconsolate expression.

Man From the Sun

"I'm leaving England. I've had enough of being the wrong colour."

Cleve in the opening scene of *Man From the Sun*.

Britain is changing as migration from the Commonwealth begins to impact the country. This is the new and pioneering theme of *Man From the Sun*. Written and produced by John Elliot and shown on 8 November, the play conveys the personal toll that racist rejection can take on one Black man (Errol John in the central role of Cleve), along with the important role of community activism in changing things for the better.

When the BBC asks the TV audience for its reaction to the play, the majority remarks that it has provoked them into greater sympathy and understanding for the new West Indian community.

Cleve (Errol John) wonders if his relationship with Maggie (Sarah Morgan) can withstand the pressures of racial prejudice.

97

The Today Programme

"What do you think politics is, if not arguing about principles?"

John Humphrys, longstanding *Today* presenter.

Sue MacGregor and James Naughtie in the *Today* studio in 1996.

Created by dynamic Talks producers Isa Benzie and Janet Quigley, on 28 October *The Today Programme* launches out of a pitched battle between the Talks and News departments. It continues to have a shaky start, wedged as it is in two 20-minute segments between a programme of hymns and Bible readings called *Lift Up Your Hearts!* and the 8.00 a.m. news bulletin.

The content is somewhat of a potpourri, too – its first show covers items such as "Briefing a Pilot at London Airport", "First Night at Liverpool", "The Sale of Napoleon's Letters", and reviews of the latest gramophone records.

John Humphrys, tenacious interrogator of *Today* for 32 years.

Mishal Husain joins the *Today* team in 2013.

Presented by…

Today's initial presenter is Alan Skempton, on Benzie's insistence that the show must have a personal voice to help define its identity. This idea really takes flight on the appointment the following year of Jack de Manio. A huge success because of his combination of informality and intense curiosity, he will be *Today*'s host until 1971.

He is joined by John Timpson in 1970, who goes on to form a partnership with Brian Redhead until 1986. The programme's reputation for exacting interviews with politicians goes back to this time; Redhead and Timpson's style has been described by John Humphrys – who takes over from Timpson – as "the cornerstone for what *The Today Programme* has become". Their natural successors are Justin Webb and Nick Robinson, who bring a naturalness and ease to the morning's national debate. It's not just men, however. The great Sue MacGregor anchors the show for 18 years, followed by Sarah Montague, Martha Kearney and Mishal Hussain.

Over the decades, *Today* will become the BBC's flagship radio programme, setting the news agenda for the day.

Test Match Special

"The bowler's Holding the batsman's Willey."

Commentator Brian Johnston's alleged commentary gaffe during the Test Match of 1976, noted by a woman listener.

Cricket has been broadcast on the BBC since radio's early years, but *Test Match Special*, or TMS as it will become known, launches on 30 May, in the hope that ball-by-ball coverage can sustain a listening audience. "Yes," is the resounding answer, and so a BBC sporting tradition is born. Continuity of producer helps it, plus an array of star commentators, including the great John Arlott, Brian Johnston and Henry Blofeld (to name but three) who exercize what cricketing bible *Wisden* calls "a disciplined line combined with a wild imagination".

Over time, other traditions linked to *TMS* emerge, notably one in which supportive listeners send home-made cakes to the commentary team. Cakes come from the great and good, too, including Prime Minister Theresa May, the Duchess of Cornwall and the Queen herself, who sends a fruitcake "baked under her close supervision".

One of the greats! Cricket commentator Brian Johnston in 1982.

The Radiophonic Workshop

"My most beautiful sound at the time was a tatty green BBC lampshade."

Delia Derbyshire talking about The Radiophonic Workshop.

**Delia Derbyshire
(1937–2001)**

Delia Derbyshire joins the BBC as a studio manager in 1960, then two years later finds her spiritual home at the Radiophonic Workshop. She stays for 11 years and creates over 200 compositions, including arranging the famous *Doctor Who* theme and many other innovative soundscapes. After the BBC, she collaborates with mainstream players such as the RSC as well as small, independent studios, always restlessly seeking acoustic innovation. Not until after her death is her contribution to contemporary music properly recognized.

The Radiophonic Workshop opens for business on 14 April from Room 13 at the BBC's Maida Vale Studios. Established to provide theme tunes, incidental music and effects for BBC programmes, the Workshop's best-known creation is the chilling *Doctor Who* theme, celebrated as one of the most iconic pieces of electronic music. The output does not stop there – it ranges across the theme tune for sci-fi series *Blake's 7* (1978) and schools' programmes such as *Look and Read* (1967), to sound effects for *The Goon Show* (1951).

The Workshop has come out of BBC Drama, where pioneer studio managers Daphne Oram and Desmond Briscoe have been pestering the BBC to experiment with new sound manipulation (or "wobbulating" as one contributor later calls it) and the modish "musique concrete". Eventually, through the Workshop, they create a collective of skilled musical innovators, including Dick Mills, Delia Derbyshire, Brian Hodgson, Paddy Kingsland, Peter Howell, John Baker and Elizabeth Parker.

The Radiophonic Workshop closes in 1998, but its influence will endure, impacting on generations of musicians and bands, from Roxy Music, The Human League and Portishead to Mizz Beats.

Composer Elizabeth Parker in
the Workshop in 1985.

Grandstand

"Well done BBC, you have done a good job for all of us outdoor sports lovers. Long overdue, but very welcome!"

Audience Research Report, 1958

David Coleman (1926–2013)

David Coleman makes his broadcast debut on *Sportsview* before becoming the face of *Grandstand* for ten years, to be followed by the lead anchor role on *Sportsnight* (later retitled *Sportsnight With Coleman*). He goes on to present *Sports Review of the Year*, as well as covering 16 Olympic Games as lead commentator. Famous for his heat-of-the-moment on-air gaffes – or "Colemanballs" as satirical magazine *Private Eye* has it, he is also universally respected as the creator of the grammar of broadcast sporting commentary.

Grandstand arrives on Saturday, 11 October, with a jaunty theme tune – "News Scoop" by Len Stevens – and an ambition to transform BBC Sports coverage.

Sportsview, launched in 1954 has begun the process, but *Grandstand* takes sports reporting to the next level, linking multiple sporting events with studio discussion and the latest gadgetry. Or, as the *Radio Times* reports, "all the machinery that goes with high-speed sports reporting – batteries of tape machines, a giant-sized scoreboard, the sub-editors' table".

The first programme is presented by the celebrated face of BBC Sport, Peter Dimmock, and features golf from St Andrews, racing from Ascot and the Horse of the Year Show from Haringey. Dimmock's original choice of presenter is a fresh-faced reporter named David Coleman. He is initially unavailable, but soon joins the show and becomes "the image and face of BBC Sport", as BBC management will soon describe him. He anchors the show until 1968, to be followed by Frank Bough, Des Lynam and Steve Rider.

Grandstand will eventually become the world's longest-running live sports programme, running from 1958 to 2007. It will close to make way for more flexible digital programming options, but still casts a long and impressive shadow.

David Coleman on the *Grandstand* set in 1978.

Blue Peter

"Here's one I made earlier..."

The famous line from every *Blue Peter* presenter, as they produce their finished "make".

Blue Peter's famous editor Biddy Baxter proudly displays the iconic *Blue Peter* badge.

In this year

21 March
Globe-trotting reporter Alan Whicker's "Whicker's World" airs on *Tonight*.

14 April
The BBC Radiophonic Workshop is established.

1 July
The first edition of radio comedy *Beyond Our Ken*.

17 July
Radio drama *The Flying Doctor* takes to the air in the Australian Outback.

10 October
Grandstand

16 October
Blue Peter

Blue Peter, the longest-running children's programme in the world – and quite possibly the best-loved – begins on 16 October.

Created by John Hunter Blair and first presented by Leila Williams and Christopher Trace, it is named after the flag a ship hoists on its maiden voyage. Fifteen minutes long and scheduled to run for only six weeks, it is transformed under the long editorship of Biddy Baxter, who listens attentively to its audience and puts their suggestions at the heart of the show.

Presenters, pets and "makes"

The show soon becomes famous for so many iconic features – from its daredevil presenters, best personified by John Noakes climbing Nelson's Column or Helen Skelton walking a high-wire between the chimneys of Battersea Power Station, to its beloved studio pets and its celebrated "makes" - cookery and craft projects made with household objects such as "chocolate bean tubes" and "detergent bottles".

Over the years, *Blue Peter* also champions charitable campaigns, inspiring children to collect stamps, cans and clothes, in order to raise money for everything from the Guide Dogs for the Blind Association to WaterAid. Participation is rewarded with a coveted *Blue Peter* badge, expanding to a themed group of eight, covering creativity, environment, music and sport, right up to the gold *Blue Peter* badge, only awarded to a very few personalities for very special achievements.

Like many long-running BBC programmes, *Blue Peter* will be endlessly parodied on comedy sketch shows; scratch the cynicism, however, and underneath is a true affection for this enduring part of so many people's childhoods.

The much-loved trio of Peter Purves, Valerie Singleton and John Noakes, with assorted pets, in 1967.

102

1958

Face to Face

Martin Luther King Jr. faces
journalist John Freeman in 1961.

Poet Edith Sitwell is revealing
about her ideals and her ambitions
in 1969.

Martin Luther King Jr., Evelyn Waugh, Stirling Moss, Edith Sitwell, Bertrand Russell, Carl Jung, Tony Hancock – these are just a few of the great names of the day interviewed by journalist John Freeman in the groundbreaking *Face to Face*, beginning on 4 February.

Interviewer and interviewee face each other in an otherwise darkened studio. Skilful, close-up camerawork focused on the interviewee's face reveals their reactions to Freeman's penetrating questions – already well-displayed from his work on *Panorama* (1953).

The first guest on *Face to Face* is the renowned criminal lawyer Lord Birkett. He proves a relaxed interviewee, unlike some that follow – notably Gilbert Harding who breaks down in tears when discussing his mother, and Tony Hancock who struggles to talk about his health and happiness.

In 1988, Freeman is himself interviewed by Dr Anthony Clare, who will create a later parallel project on Radio 4: his famous *In the Psychiatrist's Chair* series (1982).

Juke Box Jury

"I think it will be a hit, mainly because I think it's terrible."

Jazz singer Carmen McRae (Roy Orbison's "Blue Angel" is the record in question), 1960.

**David Jacobs
(1926–2013)**

David Jacobs is the evergreen radio and TV charmer. A highly popular host of *Juke Box Jury* (1957–69), he then becomes one of the original presenters of *Top of the Pops* in 1964, until making a change of gear in 1967 to handle the more contentious issues of popular topical debate programme *Any Questions?* He returns to music radio in later years, as the consummate host of a variety of Radio 2 easy-listening music shows, including *The David Jacobs Collection.*

David Jacobs spins the discs for his jury of Chris Denning, Penny Valentine, Mel Tormé and Janette Scott, in 1967.

Pop music goes mainstream with *Juke Box Jury* on 1 June, achieving, at its peak, an audience of 12 million. Each week, host David Jacobs plays a selection of singles on a large juke box to a panel of four celebrities. The disc starts to play and a roving camera pans over their faces – as well as those of the teenage audience – seeking reactions as they listen. And the big question is: will they judge the record a future "Hit" or "Miss"? The jury's verdict is greeted by the host sounding a joyful bell or a dismissive klaxon.

A range of celebrity judges appears on the show, including pop stars such as The Beatles (pandemonium ensues!) as well as the more unlikely comic Charlie Drake and comedy actor Fenella Fielding. But that has the desired effect of bringing in a wide audience of what the *Radio Times* describes as "the fans and the frankly fascinated".

The show's most potentially embarrassing moment is the appearance of a "mystery guest", who has to face the jury once they have revealed their Hit! – or Miss! – judgement. He or she has to literally face the music!

Election Night Excitement

"How did people know who to vote for before they had TV, dear?"

Caption to Vicky cartoon in the *Daily Mirror*, February 1958.

Grace Wyndham Goldie
(1900–1986)

Often called "The First Lady of Television', Grace Wyndham Goldie begins her BBC career as a critic on *The Listener*, before rapidly rising through radio Talks to TV Talks, and eventually to Head of Talks and Current Affairs. This is an amazing achievement in the male-dominated world of the BBC of the day. Once in control, she forges the flagship series *Panorama*, creates the BBC's first coverage of national elections, and recruits a team of new TV talent for the future.

In 1955, the BBC dipped its toe into the formal waters of election coverage. Now nearly five years later, and on the brink of a much more uncertain election result, the BBC ramps up its coverage. And, of course, there is now real competition for the first time from ITV.

An initial difference is the creation of BBC Hustings, a UK-wide strand of lead-up coverage, with independent chairmen posing audience questions to the prospective candidates.

On election night itself, there is a much more ample and dynamic service, as over 50 cameras are at work around the country, reporting results minute by minute and sending them back to a specially equipped studio at the BBC's Lime Grove. On the TV set, a new battery of charts, diagrams and the soon-to-be-famous "swingometer"

Presenter Cliff Michelmore discusses the results with successful MP candidate Geoffrey Johnson-Smith.

appears showing how many seats a party needs to win. Meanwhile, in the background an "electronic brain" (early computer) processes the results and attempts to predict the eventual outcome.

Of a potential audience of 22 million, 13 million watch the BBC, 5.5 million watch ITV. The result? A Conservative victory for new PM Harold MacMillan. Writing in the BBC's internal magazine, Grace Wyndham Goldie, the architect of the Corporation's election coverage, stresses the new skills and services required to make a success of this election, using "people of a wide range of experience gained in other fields: commentators, engineers, caption artists, producers, newsmen".

> "The fastest and most comprehensive coverage."
>
> Radio Times

The Election Night studio, presided over by Richard Dimbleby, with the first "swingometer" on the wall behind him.

The Beatles perform
"Paperback Writer" on
Top of the Pops in 1966.

The 1960s
Pop Goes the BBC

Deference begins its disappearing trick, as satire booms and teenagers claim their music, moves and dedicated airtime. It is only a matter of time before we bid goodbye to black-and-white days and switch onto colour. Yes, the Wimbledon grass really is green!

The Opening of BBC Television Centre

"What Hollywood is to the film industry, the new Television Centre in London will become to television."

Manchester Evening Chronicle

Television Centre lights up for business in 1960.

Designed by architect Graham Dawbarn, Television Centre opens on 29 January 1960 as one of the most technically advanced TV production hubs in the world. It is the world's first purpose-built television centre, and the third to open – after ABC in Australia and Granada Television in Manchester.

Its distinctive circular design has a quizzical beginning. The story goes that Dawbarn, having received a doorstep of a brief from the BBC commissioners, was sitting in the pub wondering how on earth to begin the project. On the back of an envelope, he doodled a question mark, then realized in a Eureka moment that he had found the perfect architectural blueprint for the new centre.

In the main block, (which will come to be affectionately known as "The Doughnut"), he decides to house the technical areas, plus facilities for artists and administration, while around it he places the TV studios with easy external access.

Notable TV programmes to come out of Television Centre will include, *Blue Peter* (from 1960), *Doctor Who* (1963), *Top of the Pops* (1964), *Monty Python's Flying Circus* (1969), *Fawlty Towers* (1975), *Absolutely Fabulous* (1992), and *Strictly Come Dancing* (2004). The building comes to symbolize "Television" to generations, who grow up watching the *Blue Peter* presenters in its famous garden (opened in 1974).

The building will be sold by the BBC in 2010 and developed as a combined media/retail/accommodation complex, although programme-making still continues in the three main studios.

The iconic circular form of the building rises up in 1957.

Architect Graham Dawbarn's question-mark doodle, as he muses on the design for Television Centre in 1949.

111

Television Centre
Clockwise from top: Early model of the building, 1950; ready for business, 1960; main reception, 1960; statue of Helios by T. B. Huxley-Jones at the centre of the building, 1960; BBC 2 cameraman, 1964; opening night in studio, 29 June 1960.

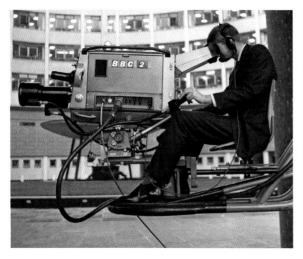

Songs of Praise

"An oasis of innocence and goodness in television schedules full of programmes about sordid people doing disgusting things to one another."

Comic actor Rowan Atkinson on the 30th anniversary of the programme in 1991

The longest-running religious television programme in the world opens its church doors on 1 October. It takes a popular radio format, *Sunday Half Hour*, and adapts it for the roving eye of the camera. The first programme comes from the Tabernacle Baptist Chapel in Cardiff, with guest soloist Heather Harper, and it sets the tone for a show that celebrates congregational singing right across the UK. Celebrity presenters, including Geoffrey Wheeler, Cliff Michelmore, Sally Magnusson, Alan Titchmarsh and Aled Jones, will add profile to the show, and it also pops up later in the comedy *The Vicar of Dibley* (1994), blurring the boundaries of real and fictional religion.

Although religious belief and churchgoing will decline in the UK post the permissive 1960s, religion is still seen as a key element of the BBC's public service remit, and *Songs of Praise* reinvents itself across the decades to maintain a prominent place in the BBC TV schedule.

Songs of Praise – a celebration of congregational singing.

Paddy (Miriam Karlin) ponders what the boss (Peter Jones) is up to this time.

In this year

10 June
The first live broadcast from the USSR.

1 October
Songs of Praise

2 October
Robert Robinson introduces *Points of View*, in which viewers write in with their comments about BBC programmes.

6 October
The Rag Trade

8 October
In Touch, the first radio programme specifically for the blind.

The Rag Trade

"Everybody out!"

Paddy's regular summons to her (female) co-workers to down tools

First broadcast on 6 October, Ronald Wolfe and Ronald Chesney's caustic comedy *The Rag Trade* is unusual for the time in that it gives all the best lines to the women characters.

With trade union unrest regularly making headline news, the sitcom, set in a clothing factory, dramatizes the regular clashes between its miserly owner (Peter Jones) and pattern cutter (Reg Varney) and its powerful female workforce. This consists of a roll call of great British female character actors, including Sheila Hancock, Esma Cannon, Wanda Ventham, Barbara Windsor, Irene Handl and Miriam Karlin (as militant shop steward Paddy).

115

Steptoe and Son

No escape for Harold (Harry H. Corbett) from his rag-and-bone man father (Wilfrid Brambell).

One of the most influential TV comedies ever written, *Steptoe and Son* is created by Galton and Simpson of *Hancock's Half Hour* fame. First broadcast on 7 June, the sitcom is an instant hit. It lasts, on and off, until 1974, spawning two feature films and a radio version.

Steptoe and Son are rag-and-bone men, inhabiting the same cluttered junkyard home every week. More importantly, they are also psychologically trapped: time and again the aspirational Harold (Harry H. Corbett) attempts to escape the clutches of his father Albert (Wilfrid Brambell), only for his hopes to be dashed. *Steptoe and Son* walks a narrow line between comedy and tragedy, as well as making pointed comments on the state of the nation, in particular class and sexual mores. It is so popular that Prime Minister Harold Wilson pressurizes the BBC to move its transmission on election day 1966, as he fears voters will stay away from the polling booth.

That Was the Week That Was

"This TV newcomer smiles as she bites."

Daily Herald TV critic Dennis Potter on the launch of TW3.

A new Director-General has arrived at the BBC, Hugh Carleton-Greene, brother of author Graham Greene, and he is determined to "open the windows and dissipate the ivory tower stuffiness". Nowhere is this more evident than in the boundary-breaking satire show *That Was the Week That Was* (*TW3*), Fronted by 23-year-old newcomer David Frost, and with a cast including Millicent Martin, Kenneth Cope, David Kernan, Roy Kinnear, Bernard Levin, Lance Percival and Willie Rushton, *TW3* takes the nation by storm. Beginning on 24 November, the live show combines songs with sketches and cartoons in a freewheeling format. The writers challenge every contemporary taboo – government, politics, church, marriage and sex – with often scandalous results. The presentation introduces a new look, with producer Ned Sherrin stripping away TV artifice to show the studio in all its behind-the-scenes glory.

TW3 will only run for two series, the rumour being it is too hot for the BBC to handle. Cancelled just before the 1964 Election year, it signals the beginning of the end of Establishment deference.

David Frost and the TW3 cast (*right to left*): Lance Percival, David Kernan, Al Mancini, Millicent Martin, Kenneth Cope, Roy Kinnear, Willie Rushton and Irwin Watson.

Doctor Who

"A kind of science-fiction James Bond with a touch of the Renaissance Man."

Jon Pertwee on the role of Doctor Who.

The first three Doctors: William Hartnell, Patrick Troughton and Jon Pertwee, 1972.

The show's opening titles accompanied by the famous theme music, 1963.

"Television history will be made on Saturday," announces *Titbits* magazine on 23 November. And so begins the longest-running TV sci-fi series in the world, albeit with gaps in its programming. Devised to fill the Saturday evening slot between *Grandstand* and *Juke Box Jury*, *Doctor Who* is the brainchild of new Head of Drama Sydney Newman (brought in from ABC) and producer Verity Lambert. It has a painful internal genesis with wrangles over budget and resources, but eventually the first show airs, introduced by its utterly original theme tune, composed by Ron Grainer and reinvented by the Radiophonic Workshop.

In essence, *Doctor Who* is the struggle of good against evil, captured in the adventures of a set of time-travellers, led by the Doctor (a non-human Time Lord) and the Doctor's companions, which at

Exterminate! The most
famous of the Doctor's
enemies are the Daleks.

various times include a robotic dog named K9. Their enemies are
many and frequent, including the notorious Daleks, created by BBC
designer Raymond Cusick who gets the idea when "fiddling with a
pepperpot".

Doctor Who?

The original doctor is classical actor William Hartnell, but he soon
gives way to a parade of characterful and different Doctors: Patrick
Troughton, Jon Pertwee, Tom Baker, Peter Davison, Colin Baker,
Sylvester McCoy, Paul McGann, Christopher Eccleston, David
Tennant, Matt Smith, Peter Capaldi … right up to the character's
transformation (or, in the language of the series, "regeneration")
into the first female Doctor, Jodie Whittaker.

The show maintains a vestige of its BBC educational
DNA, through its frequent trips back into history, from Marco
Polo to Queen Victoria, as well as reflecting each decade's
underlying concerns, including the Bomb, dictator domination and
shifting patterns of class and gender.

It will be gloriously rebooted in 2005 by screenwriter and *Doctor
Who* "superfan", Russell T. Davies, becoming one of the BBC's most
globally successful brands.

The first female Doctor
(Jodie Whitaker) does battle,
supported by Graham (Bradley
Walsh), Yaz (Mandip Gill) and
Ryan (Tosin Cole), 2019.

In this year

30 September
The BBC-TV globe emblem
appears between
programmes, featuring a
continuity announcer.

23 November
Doctor Who

Top of the Pops

"It's Number One… It's *Top of the Pops*!"

Cue for the last record to play on the show each week.

Just a few of the DJs who front the most popular pop show on TV (*from top*): Alan Freeman, David "Kid" Jensen and Janice Long.

Times are a-changing, and nowhere is this more apparent than in the rising passion for pop music. In 1957, the BBC creates *Six-Five Special* in 1957, "designed for the young in spirit" (*Radio Times*). It features rock'n'roll, skiffle and jazz, and quickly proves popular. But with increasing pressure from ITV pop shows like *Thank Your Lucky Stars* (1961) and *Ready Steady Go!* (1963), the BBC launches its most successful pop juggernaut ever: *Top of the Pops*. Beginning on Wednesday, 1 January 1964, it will last and last…

"Why does a 'pop' come to the top?" quizzes *Radio Times* in its tie-in issue. The show answers this question by presenting chart-topping singles as well as songs on their way up the Top 20. Its first presenter is DJ Jimmy Savile; the Rolling Stones open the show with "I Wanna Be Your Man" and acts featured include Dusty Springfield, The Dave Clark Five, The Hollies, and The Swinging Blue Jeans. The Beatles' "I Want to Hold Your Hand" is this week's Number One.

A pop institution

Top of the Pops quickly makes itself unmissable – the shop window for any artist or band hoping for mainstream success, mainly because, with the exception of *Ready Steady Go!*, which finishes in December 1966, there is no other TV show to compete with it. And the delight of seeing the top singers of the day – even though they initially mime to their songs – along with the suspense-filled, end-of-show countdown to the week's top-selling single, ensures a huge youth following.

Its troupe of dancers interpreting a chart hit – initially called the Go-Jos, then mutating over time to Pan's People, Ruby Flipper, Legs and Co. and finally Zoo – are also a big and distinctive draw. Dancer Dee Dee later recalls: "It was electric and thrilling… we all had a great time in the 60s and 70s."

The Kinks on *Top of the Pops*, the year the show opens.

120

1964

The Launch of BBC Two

"I ploughed on through every scrap of unedited Reuters tape
they could feed me."

Gerald Priestland on the horrors of having to hold the fort on air during the ill-fated BBC Two launch.

In this year

1 January
Top of the Pops

3 April
Madcap radio comedy *I'm Sorry I'll Read That Again* launches the careers of future comedy stars Tim Brooke-Taylor, John Cleese, Graeme Garden and Bill Oddie.

20 April
BBC2 (later renamed BBC Two) launches.

21 April
Children's show *Play School* airs on BBC2.

2 May
Science documentary series *Horizon*.

22 August
Kenneth Wolstenholme presents *Match of the Day*.

28 October
The first *Wednesday Play:* Jean-Paul Sartre's *In Camera*.

16 December
Rodney Bewes and James Bolam star as Bob and Terry in Dick Clement and Ian La Frenais' sitcom *The Likely Lads*.

Following a government review of the state of broadcasting – and an agreed expansion – the BBC launches its second channel on 20 April 1964. Or rather, it doesn't!

A fire at Battersea Power Station creates a blackout, and West London goes dark. The ambitious entertainment schedule, featuring live comedy from The Alberts, a performance of Cole Porter's musical *Kiss Me Kate* with American star Howard Keel, comedy from "Soviet Union's leading comedian" Arkady Raikin and a grand fireworks display, disappears. Instead, the only programming is a short and panic-stricken news announcement from Gerald Priestland in Alexandra Palace, which was unaffected by the power cut.

The first full broadcast to go out on BBC Two is a new programme for pre-school children, *Play School*, transmitted at 11a.m. the following morning, with the opening night line-up eventually shown that evening.

The new channel is transmitted on 625 lines, rather than BBC One's 405 lines. This aligns the UK with Europe, and gives better picture quality, but does demand the purchase of a new TV set and aerial.

Eventually, BBC Two settles down, earning itself a reputation for alternative and innovative programming, from *The Great War* to *Jazz 625*. In future years, it will also pioneer new colour output and an increased package of educational programming, notably the ground-breaking Open University collaboration in 1971.

Hopeful on-screen announcement
to cover the failed launch.

Vision On

"Show them, don't tell them."

Tony Hart

The magic combination of presenters Tony Hart and Pat Keysell.

Having created a monthly show called *For Deaf Children* in 1952, pioneering Children's TV producer Ursula Eason is keen to do more for deaf and hearing children combined. On 6 March this year she launches *Vision On*. Bursting with creative ideas and formats, the show becomes a massive hit with its young audience, and before very long the programme becomes just as popular with its hearing as with its non-hearing audience.

It has two charming presenters – actor/teacher Pat Keysell, who is joined in the second series by artist Tony Hart. With his quick-as-a-flash drawing skills, he encourages every child to have a go at being creative, and entries to the show's Gallery come flooding in (up to 8,000 per week). Later, the show also becomes known for its surreal films and animations, developed by the creative team eventually to become the globally successful Aardman Animations.

Tomorrow's World and Horizon

"The original *Tomorrow's World* inspired a generation –
it certainly inspired me back in the 1970s."

Professor Brian Cox

A new passion for science emerges in the 1960s, underpinned by Prime Minister Harold Wilson's vision of a new Britain, "forged out of the white heat of technology". New broadcasting feeds and supports this ideal, nowhere more apparent than in two long-running shows, *Tomorrow's World*, which airs on 7 July this year attracting a mass audience, and *Horizon*, aimed at adult viewers, which began on May 2 the previous year.

Raymond Baxter and Bob Symes examine a model of the new advanced passenger train, 1969.

The world of the future

Initially presented by ex-RAF pilot Raymond Baxter – to be followed by a succession of illustrious presenters, including James Burke, Michael Rodd, Judith Hahn, Maggie Philbin, Howard Stableford and Peter Snow – *Tomorrow's World* imagines our future for us by showcasing the inventions that will shape it.

124

Peter Snow and Philippa Forrester grapple with tomorrow's world in 1999.

For the first time we will see a computer terminal in our home and workplace (1967), money accessible via an ATM (1969), pop music created by machines – famously demonstrated by the German band Kraftwerk (1974), a phone that is mobile (1979) and the dawn of the Information Superhighway (1994).

Of course, not all the inventions are realized, and some of the show's live demonstrations are famously flawed – audiences will later remember the innovative car jack that collapses! Presenters have to learn how to cope with these on-air difficulties, ripe for parody in contemporary comedy shows such as *Not the Nine O'Clock News* (1979). But *Tomorrow's World*'s role in helping us imagine the future before us is undeniable.

Viewing the universe

Horizon provides "a platform from which some of the world's greatest scientists and philosophers can communicate their curiosity, observations, and reflections, and infuse into our common knowledge their changing views of the universe". So writes the programme's producer, Philip Daly, in the *Radio Times*.

In 2006, *Horizon* will explore the impact of a killer pandemic with amazing prescience.

Horizon will go on to produce many controversial and influential programmes, from its 1972 programme exposing the horrors of commercial whaling to its ground-breaking 1983 documentary about AIDS, the first on the subject in the UK. It will continue, unpacking the scientific issues – large and small – that touch our everyday lives.

The Magic Roundabout, Jackanory and Play School

"I'll tell you a story about Jackanory. And now my story's begun."

Opening words of the long-running *Jackanory* series.

In this year

7 January
Not Only... But Also: groundbreaking comedy starring Peter Cook and Dudley Moore.

7 March
Radio comedy *Round the Horne*, hosted by Kenneth Horne and written by Barry Took and Marty Feldman.

31 March
Expert Arthur Negus values antiques in the quiz show *Going for a Song*.

30 May
The World of Wooster brings author P. G. Wodehouse's comic creations Jeeves and Bertie Wooster to TV.

7 July
Tomorrows World

4 October
Radio news and current affairs programme *The World at One*.

18 October
The Magic Roundabout

13 December
Jackanory

Join Florence and her colourful friends on the entrancing Magic Roundabout.

The Magic Roundabout begins its first entrancing rotation on 18 October this year. Eric Thompson brilliantly reinvents the French original, *Le Manège Enchanté*, and creates an instant TV children's classic – though adults fall in love with it, too.

It's the story of Mr Rusty's merry-go-round, abandoned by the children until a magic jack-in-the box brings them back. Its memorable characters include the little girl Florence, Dougal the dog (apparently based on Tony Hancock), Brian the snail, Ermintrude the cow, Dylan the rabbit, and Mr McHenry the roundabout operator. Generating a feature film, *Dougal and the Blue Cat*, in 1972, the much-loved show runs until 1977.

**Floella Benjamin
(1949–**

Originally from Trinidad, Benjamin makes her name in the groundbreaking *Play School* and *Play Away* from the late 1970s. As one of its first Black presenters, Floella changes the perceptions of generations of viewers.

Following more than a decade on air, she creates her own TV production company in 1987, making children's programmes for a range of channels, and winning a BAFTA Special Lifetime Achievement Award in 2004. Recognized for her advocacy work for children, she will eventually become Baroness Benjamin of Beckenham.

Jackanory

This classic storytelling show begins to cast its spell on 13 December 1965, with "Cap of Rushes", told by Lee Montague. Many famous actors take on the storyteller role after him, including Kenneth Williams, Geraldine McEwan, Alan Bennett, Michael Hordern, Rik Mayall, Judi Dench and Tony Robinson. Top billing goes to Bernard Cribbins who presents 111 episodes, far more than anyone else.

Jackanory persuades generations of reluctant readers to pick up a book. Concluding in 1996, the programme spawns the spin-off *Jackanory Playhouse* (1972) and *Jackanory Junior* (2007), and is reinvented for younger children in the enduring CBeebies *Bedtime Stories* with its host of celebrity readers.

Play School

"What window are we going to go through today?" This is the regular *Play School* invitation, as the programme takes pre-school children on a journey of discovery through one of its on-set magic openings.

Launched the previous year, on 21 April 1964, and by accident the first proper programme on BBC Two, the show uses "all the advantages of television to do the job of a nursery school in its own exciting way' (producer Joy Whitby writing in *Radio Times*). Presenters over the years include the charismatic Brian Cant, Toni Arthur, Carol Chell, Floella Benjamin and Johnny Ball. To be joined by the toys Humpty, Jemima, Hamble, Big Ted and Little Ted and Poppy.

The ever-popular Bernard Cribbins reads Tolkein's *The Hobbit* on Jackanory, 1979.

The World Cup Final

"And here comes Hurst, he's got... Some people are on the pitch!
They think it's all over! It is now! It's four!"

Match commentator Kenneth Wolstenholme's immortal words as Geoff Hurst scores England's decisive winning goal.

World Cup commentator Kenneth Wolstenholme knows "It's all over", as he utters the famous words.

It is probably the biggest event in British sporting history. The live broadcast of England's World Cup victory over West Germany (4 goals to 2) captivates the nation on 30 July – as over 32 million people watch. It will be talked about for years to come.

The BBC and competing broadcaster ITV come together for the World Cup, to ensure viewers get the best and most detailed coverage. BBC Sports anchor David Coleman previews the match on air from noon, while match commentary is delivered by Kenneth Wolstenholme, whose words at the end of the game will gain him broadcast immortality.

The innovation of "action replays" – the term is created by the BBC's own Head of Sport Bryan Cowgill – are also seen for the first time during this match.

Ninety-three thousand spectators cram into Wembley Stadium to watch the game, including the Queen and Prince Philip. It is a tense match, going into extra time, until eventually Captain Bobby Moore holds the gold-plated Jules Rimet trophy aloft, following a hat-trick by striker Geoff Hurst.

BBC cameras record the thrilling moment of English victory. Football will continue to have a magnetic appeal for TV audiences. Over 25 million watch England's defeat to Germany in the 1990 FIFA World Cup semi-final, and almost 31 million watch the 2020 Euro Final where England lose in extra time to Italy.

Match of the Day

> "I've only got a Saturday job, so my weekdays
> are generally pretty free."
>
> *Match of the Day* presenter Gary Lineker (from 1999)

In this year

5 January
Police drama *Softly Softly*.

2 February
Hard-hitting documentary series *Man Alive*.

6 June
Sitcom *Till Death Us Do Part*, written by Johnny Speight, introduces viewers to bigoted Alf Garnett and his long-suffering family.

30 July
England's World Cup victory is televised.

1 September
The first edition of Dusty Springfield's own show, *Dusty*, airs.

16 November
Controversial TV play *Cathy Come Home* broadcast.

28 December
Director Jonathan Miller presents his star-studded interpretation of Lewis Carroll's *Alice in Wonderland*.

The first regular football programme on television – to become the longest-running football programme in the world – Match of the Day begins some two years earlier on 22 August 1964. It arrives at the start of the 1964–65 season, but the identity of the match – Liverpool v. Arsenal – is kept secret until 4.00 p.m., by agreement with the Football League who fear that fans will stay away if they know the match is to appear on television.

The show is transmitted on the new channel BBC Two, giving it a small audience but much higher definition picture (625 lines). The programme is introduced by Major Leslie Statham's tune "Drum Majorette"; this will be replaced by Barry Stoller's iconic gladiatorial theme in 1970.

Match of the Day helps prepare the production crew for the 1966 World Cup, and its popularity soon prompts a move to BBC One, where it remains a mainstay of the schedule, renowned for its expert commentators and ex-pro pundits.

Match of the Day presenter Jimmy Hill on set in 1986.

The UK Wins Eurovision

*"A lot of writers made the mistake of writing for Sandie Shaw,
while we wrote for Europe… You have three minutes for a Eurovision song
and the meter's running."*

Phil Coulter, composer of "Puppet on a String".

Sandie Shaw actually hates the song but "Puppet on a String" not only wins Eurovision, it tops the UK singles chart, giving her a third Number One, a record for a female artist at that time.

Created in 1955, the European Broadcasting Union (EBU) decides to initiate a pan-Europe song contest as an exercise in post-war harmony. The BBC misses the first competition that year, but is ready for the next one in 1956, as well as helping to build up the competition's television appeal and stature. In fact, its name is actually coined by BBC publicist George Campey, who proffers it as a snappier alternative to "Continental Television Exchange".

Puppet on a String
The UK entry for Eurovision 1967, held in Vienna, is "Puppet on a String", performed by a characteristically barefoot Sandie Shaw. Penned by Bill Martin and Phil Coulter, the song is famous for being the UK's first winner of the contest. Once heard, never forgotten! In 1968, London hosts Eurovision, in colour for the first time. The contest will grow to become the world's biggest live music event.

The Launch of Radio 1

*"I woke up early specially, with my little brown radio under the bedcovers,
feeling literally sick with anticipation – a new station for us kids!!!"*

Young Radio 1 listener (aged 13), 1967.

On 30 September 1967 it's all change for the BBC! Responding (a little late) to the zeitgeist, the Corporation launches its first-ever radio pop station, and reorganizes the legacy stations of World War II into three new radio networks.

Happy 1st Birthday Radio 1! Controller of Radios 1 and 2, Robin Scott holds the cake for the station's DJs in 1968.

Wider Radio

In parallel with the birth of Radio 1, the old Light Programme, Third Programme and Home Service are replaced by Radios 2, 3 and 4. The old names are still used for two years, while listeners get used to the new ones. Developments from the late 1980s onwards will include the specialist networks 1Xtra, 4 Extra, 5 Live, 6 Music and Asian Network, catering for new and different audiences.

Revolution has been in the air for a while, with pirate radio stations – in particular Radio Caroline and Radio London – attracting a huge youth following and denting the BBC's potential audience. They are able to do so thanks to a loophole in the law, by broadcasting from international waters. Then, at a stroke, the Labour government's Marine Broadcasting Offences Act makes pirate radio illegal, and the BBC invites the pirates on board the good ship Broadcasting House. DJs such as Tony Blackburn, John Peel, Kenny Everett, Emperor Rosko, Johnnie Walker, Ed "Stewpot" Stewart, Dave Lee Travis, Simon Dee and Tommy Vance soon become household names.

It's 7.00 a.m., and after a jingle "The voice of Radio 1 – just for fun", ex-Radio Caroline DJ Tony Blackburn welcomes listeners to the new network and spins his first disc: "Flowers in the Rain" by The Move. The old radio guard mutters, while young listeners embrace the new, improvised and easy Radio 1 style and the latest pop sounds with zestful enthusiasm.

The Arrival of Colour Television

"Green, green, the grass is really green!"

Excited viewer on seeing Wimbledon tennis in colour for the first time in 1967.

Experimentation with colour TV goes right back to the 1920s and 30s with the pioneering work of John Logie Baird. It begins in earnest at the BBC in the mid 1950s, but is complicated by multiple broadcast formats and standards. Nevertheless, David Attenborough, then Controller of BBC Two, decides to push forward the launch of colour for his channel, so he can claim it as the first in Europe.

Lack of studio facilities means that most of the early colour content is either outside broadcasts or films. Attenborough decides to use the Wimbledon Tennis Championships as a piece of star content in colour to motivate audiences to go out and buy the required new TV sets and licences. Later, he has a surprise hit with *Pot Black* (23 July 1969), one of the first times snooker has appeared on British TV screens, which uses colour to obvious visual effect.

No need to tell viewers where the blue is any more – the advent of colour really brings snooker to life.

132

A colour camera surveys the court at Wimbledon.

In this year

1 July
The Wimbledon Championships is the first BBC Two colour broadcast.

2 July:
Test Card F, aka the "Bubbles" colour test card, is shown.

30 September:
Pop music station Radio 1 begins broadcasting.

1 October
Radio show *Pick of the Pops* features a rundown of the UK's Top 20 hits.

13 October
Arts show *Omnibus* airs on BBC One.

8 November
The first local radio station, Radio Leicester, opens.

22 December
Chopin's "Minute Waltz" introduces radio comedy panel show *Just a Minute*, created by Ian Messiter and presented by Nicholas Parsons.

Wimbledon Tennis

The Wimbledon Tennis Championships of 1 July 1967, shown only on BBC Two, mark the beginning of regular colour television in Britain. Attenborough announces that the channel will initially broadcast only about 5 hours a week in colour, but he is justifiably elated at his achievement in pulling off the first colour TV service in Europe: "I was as proud as a peacock!" he later exclaims.

By December, 80 per cent of BBC Two programmes are in colour. BBC One (and ITV) do not transition to colour until 15 November 1969. But this phased approach is actually helpful, as it enables the BBC to get properly kitted up and allows TV manufacturers and the general public to adapt to the new technology.

Dad's Army

"Would you all mind terribly falling in, please? Thank you so much."

Sergeant Wilson's "command" to the platoon.

Dad's Army proves so lastingly popular it is
serially adapted for radio, stage and film.

David Croft (1922–2011) and Jimmy Perry (1923–2016)

The duo's first big collaboration is *Dad's Army*,
based on Perry's experiences as a teenage
volunteer prior to conscription call-up. Success
follows, often mining their own life
experiences: the army entertainment corps in
It Ain't Half Hot, Mum (1974), and work in
Butlin's holiday camp in *Hi-de-Hi* (1980).
Croft also collaborates with Jeremy Lloyd, on
the sitcoms *Are You Being Served?* (1972) and
'Allo 'Allo (1982).

On 31 July 1968, viewers of BBC One meet the brave
and baffled members of the Walmington-on-Sea Home
Guard in *Dad's Army*, as they prepare to repel Nazi
invaders during World War II.

Its tone of ironic nostalgia is set by its jaunty theme
tune: "Who Do You Think You Are Kidding, Mr
Hitler?" sung by 1940s comedian Bud Flanagan. Its cast
of characters stems from the experienced pens of Jimmy
Perry and David Croft, and is realized by top comedy
actors. Arthur Lowe is pompous bank manager Captain
Mainwaring; John Le Mesurier is Sergeant Wilson;
Clive Dunn is gung-ho local butcher Corporal
"Jonesy" Jones; John Laurie is doom–mongering
Scottish undertaker Private Frazer; James Beck is
black-marketeer Private Walker; Arnold Ridley is
elderly Private Godfrey; and Ian Lavender is young
Private Pike (regularly lambasted by Mainwaring as
"You Stupid Boy!"). *Dad's Army* ends – but never really
ends as it continues to be broadcast on various TV
channels for decades – in 1977 after 80 episodes.

134

The Morecambe & Wise Show

"I am playing all the right notes, just not necessarily in the right order."

Eric Morecambe's famous put-down to André Previn after playing chopsticks as Grieg's Piano Concerto.

Eddie Braben (1930–2013)

In the 1950s, Eddie Braben begins writing scripts for fellow Liverpudlian Ken Dodd, before transferring to Morecambe and Wise. He is not a fan in the beginning, but becomes "the third man" of the duo, writing much of the material for their celebrated Christmas shows. The responsibility and sheer effort takes a toll on his health: "When you realized there were 20 to 25 million people looking over your shoulder all saying 'make me laugh'."

In this year

5 January
Ken Burras presents *Gardeners' World* from Oxford Botanical Gardens.

30 January
Cilla Black presents her own show, *Cilla*; theme tune "Step Inside, Love" is penned by Paul McCartney.

31 July:
Dad's Army

2 September
The Morecambe & Wise Show

On 2 September 1968, Eric Morecambe ("tall with glasses") and Ernie Wise ("short, fat hairy legs") bring their comedy show back to the BBC, having been away for a decade or so on ITV. They soon become the nation's most popular double act, culminating in the 1977 Christmas special seen by 28 million viewers, the all-time record for a comedy programme in British television history.

The first *Morecambe & Wise Show* features singer Georgia Brown and Cuban vocal group Los Zafiros as guests. The show soon attracts some of the biggest names in showbusiness, all happy to be ridiculed – from Des O'Connor and Peter Cushing (always asking for his fee) to Vanessa Redgrave and Glenda Jackson. Top musicians and singers join the queue, too – most famously André Previn (Eric insists on calling him André Preview), along with newscasters and weather forecasters abandoning their professional roles to join in the comic mayhem.

A baffled André Previn looks on as Eric and Ernie famously massacre Grieg's Piano Concerto, 1971.

The Moon Landing

"The most historic journey in the history of man, perhaps…"

Cliff Michelmore, main TV presenter for the BBC's Moon Landing coverage.

The resilient all-night studio team of James Burke, Cliff Michelmore and Patrick Moore.

On 20 July a person walks on the moon for the first time in history, and broadcasting is there to show the moment as it happens. It is extraordinary, unbelievable, and watched by 22 million people in the UK and 650 million worldwide. Technologically it is a feat, too, as there are limited satellites available to beam the pictures from the USA across the world.

In all, the BBC hosts 27 hours of coverage, including – for the first time ever – an all-night session around the Landing itself. The BBC presenting team is led by the familiar face of Cliff Michelmore, with *Tomorrow's World* expert James Burke covering the science and Patrick Moore as the fount of astronomical knowledge. "No one knew what we were looking at except Patrick," says James Burke later.

Most of the UK sees the Moon Landing in black and white, as colour will not arrive on BBC One till November of 1969, although images are available in colour on BBC Two.

Cannily, the BBC ties in the launch of the American sci-fi series *Star Trek* (soon to become a national favourite) with the Moon Landing. *Star Trek*'s first episode is called appropriately "Where No Man Has Gone Before". Fact and fiction blur as the nation becomes gripped by space fever. The legacy of the Moon Landing at the end of this decade is an extraordinary sense of global optimism: it seems we can do anything, go anywhere…

"A giant step for mankind…" Television shows the epoch-defining moment.

The Clangers

"Where are you today, then? Here, or planet of the Clangers?"

DCI Gene Hunt (Philip Glenister) ironically references *The Clangers* in *Life on Mars* (2006).

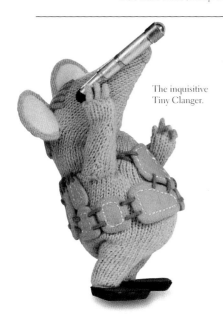

The inquisitive Tiny Clanger.

Inspired by the space race comes *The Clangers*, created by the unfailingly original Oliver Postgate and Peter Firmin – whose earlier children's successes include *Noggin the Nog* (1959) and *Ivor the Engine* (1959).

The Clangers are a family of whistling, knitted, mouse-like creatures who live on a small planet very like the moon. They share their world with the Soup Dragon – who provides them with food from the soup wells – and the Froglets, as well as being visited by alien lifeforms such as the Iron Chicken and the musical Hoots. The gentle, probing narration is provided by Oliver Postgate himself.

The Clangers will go on to delight generations of children, returning again in 2015 created by a new team that will include Postgate's son Daniel.

Monty Python's Flying Circus

"This is an ex-parrot!"

John Cleese's exasperated customer explains to pet shop owner (Michael Palin) that the parrot he has bought is, err… dead.

The Pythons (*left to right*): Terry Jones, Graham Chapman, John Cleese, Eric Idle, Terry Gilliam and Michael Palin.

Eric Idle insists in vain that he's not wearing a toupée in the "Toupée" sketch.

"And now for something completely different"… followed by the animated image of a giant naked foot squashing the life out of everything. This is the opening announcement and titles of this anarchic and surreal comedy that begins on 5 October at 11p.m. and signals the end of the 1960s comic world as we know it.

Graham Chapman, John Cleese, Eric Idle, Terry Jones, Michael Palin and animator Terry Gilliam are the core Pythons, abetted in some sketches by Carol Cleveland. The team is brought together by comedy writer, presenter and comedian Barry Took. The Pythons' first show sets the tone of its risk-taking humour, featuring "Famous Deaths" presented by Mozart, the writing of the funniest and deadliest joke in the world, and an interview with Arthur "Two Sheds" Jackson. Later will follow such famous sketches as "The Dead Parrot", "The Ministry of Silly Walks", and "The Batley Townswomen's Guild Present the Battle of Pearl Harbor".

In this year

23 February
Civilisation: A Personal View by Kenneth Clark reveals the history of Western art in colour.

21 June
The documentary *Royal Family* reveals the day-to-day life of the Queen's family.

18 July
Carla Lane and Myra Taylor's sitcom *The Liver Birds* views the romantic misadventures of two young women in Liverpool.

21 July
The Moon Landing

23 July
Pot Black

29 July
Anticipating the reality TV of later decades, Nigel Kneale's *The Year of the Sex Olympics* presents a dystopia in which an amoral elite seeks to control the masses with vicarious TV thrills.

9 September
Early evening news and current affairs show *Nationwide* begins. Presenters include Frank Bough, Sue Lawley, Bob Wellings and Michal Barratt.

5 October
Monty Python's Flying Circus

17 November
Take Three Girls tells the stories of three very different young women sharing a flat in "Swinging London".

Terry Gilliam's surreal animations define the anarchic world of *Monty Python*.

After three series, John Cleese leaves. The fourth and final series in 1974 is simply known as Monty Python. The show gains a huge following and spawns four feature films, books, records and stage shows. It also creates a new adjective in the English language: "Pythonesque", in recognition of its unparalleled uniqueness.

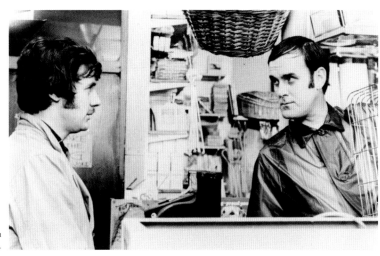

Michael Palin and John Cleese in the "Dead Parrot" sketch.

Esther Rantzen and the *That's Life!* presenting team hit the supermarket in 1983.

The 1970s
That's Life!

Reality takes centre stage with hard-hitting dramas and the first fly-on-the wall documentary. And we are all consumers now, tackling our rights with new determination – though some of us are still waiting to be served or suffering the indignities of the worst hotel in town…

Dramatizing History: The Six Wives of Henry VIII and Elizabeth R

"Developing a character over this enormous width of canvas was a challenge I just couldn't resist."

Glenda Jackson talking about *Elizabeth R* in the *Radio Times*, 1971

Keith Michell in a career-defining role as Henry VIII..

By the end of the 1960s, colour television has well and truly arrived. It is a natural showcase for the opulent costumes and pageantry of historical drama, and the decade kicks off with two hugely popular examples of the genre.

The Six Wives of Henry VIII begins on 1 January, its opening programme devoted to the story of Catherine of Aragon (Annette Crosbie) and her tumultuous divorce from the king. To be followed by Dorothy Tutin as Anne Boleyn; Anne Stallybrass as Jane Seymour; Elvi Hale as Anne of Cleves; Angela Pleasence as Catherine Howard; and Rosalie Crutchley as Catherine Parr.

All the cast are flamboyantly dressed by ingenious costume designer John Bloomfield, using painted cheap fabric, glass and even household washers to achieve Tudor period style on a BBC budget.

At the show's centre is a star performance from Keith Michell as Henry, ageing from athletic young prince to obese, gout-ridden tyrant, yet revealing a more complex, sympathetic side of the monarch than hitherto. He will repeat the role in a 1972 feature film, directed by Waris Hussein, and is one in a long line of on-screen Henrys as television tells the Tudors' story again and again…

Making history live

So successful is *The Six Wives of Henry VIII* that the BBC quickly follows it up in February 1971 with *Elizabeth R*, starring Glenda Jackson (later to reinvent herself as a real-world politician) in the title role. Over six 90-minute episodes, Jackson has to age, like Michell, from feisty youth to raddled old age, achieving this by lengthy and

painstaking make-up and wig sessions. Her portrayal of Elizabeth I is a triumph in its mix of vulnerability and steely determination, and wins many awards.

The remarkable costume design for *Elizabeth R* is by Elizabeth Waller, and, as with *The Six Wives of Henry VIII*, the costumes embark on a nation-wide tour and are the source of huge audience interest.

Both of these projects establish the BBC as the pre-eminent maker of costume drama, both in the UK and overseas, and intensify the small screen's unceasing fascination with this turbulent period of British history.

Glenda Jackson as the young Princess Elizabeth.

Jackson undergoes lengthy make-up sessions to become the older Queen Elizabeth.

Play for Today

"Television is the true national theatre."

Dennis Potter, playwright and screenwriter, writing about *Play for Today* in the *Radio Times*, 1970

Carol White as Cathy on the downward spiral to homelessness in *Cathy Come Home*.

The Wednesday Play, first broadcast on 28 October 1964 (finishes 1970), begins the strand of single dramas that, in the words of its creator, Head of Drama Sydney Newman, dramatize "the turning points in contemporary Britain". *Cathy Come Home* (1966), written by Jeremy Sandford and directed by Ken Loach, is perhaps the most famous of the Wednesday Plays, revealing the shocking impact of homelessness on a young couple and changing public perceptions almost overnight of the issue.

On 15 October this year, *The Wednesday Play* becomes *Play for Today*, continuing the same focus on sharp, contemporary writing and relevant drama. Its first play is *The Long Distance Piano Player* by Alan Sharp, starring musician Ray Davies of The Kinks. The strand goes on to showcase over 300 plays, featuring the best of new writing by the likes of John Osborne, Trevor Griffiths, Ingmar Bergman and Dennis Potter – as well as directing and acting talent. Its range is extraordinary – from social realism and comedy to sci-fi and the overtly experimental. It constantly breaks taboos and pushes boundaries, as well as frequently shocking the TV audience.

Play for Today classics include: *Edna the Inebriate Woman* (1971) by Jeremy Sandford; *Bar Mitzvah Boy* (1976) and *Spend, Spend, Spend* (1977) by Jack Rosenthal; *Abigail's Party* (1977) by Mike Leigh; *Licking Hitler* (1978) by David Hare; and *Blue Remembered Hills* (1979) by Dennis Potter. Two plays cause huge controversy and are banned: Dennis Potter's *Brimstone and Treacle* (1976, not aired until 1987) and Roy Minton's *Scum* (1977, not aired until 1991). *Play for Today* ends in 1984, as BBC drama shifts from commisioning single plays towards series and serials.

Beverly and Laurence (Alison Steadman and Tim Stern) with Angela and Tony (Janine Duvitski and John Salthouse) in *Abigail's Party*.

The Goodies

"The Goodies were completely indestructible, weren't they? A live-action cartoon mixed in with these incredible surreal flights of fancy."

Comedian Jon Culshaw

In this year

1 January
The Six Wives of Henry VIII

5 January
David Vine presents TV quiz *A Question of Sport*, featuring leading sports personalities.

23 March
Frankie Howerd stars in risqué sitcom *Up Pompeii*.

6 April
Discussion programme *Start the Week* airs on Radio 4 on Monday mornings.

6 August
The first same sex kiss on TV occurs in an adaptation of Christopher Marlowe's *Edward II*.

5 October
Consumer programme *You and Yours* airs on Radio 4.

8 November
The Goodies

"The Goodies are coming for you and you and you"… chants their opening titles. And the three of them – Tim Brooke-Taylor, Graeme Garden, Bill Oddie – arrive on their wobbly comedy "trandem" on 8 November, bringing with them their brand of much-loved cartoon humour, slapstick visuals and surreal diversions (remember the attack of Kitten Kong on the Post Office Tower?!).

They worked together earlier on madcap radio comedy *I'm Sorry, I'll Read That Again* (1964) and TV sketch show *Broaden Your Mind* (1968), but *The Goodies* expands their comic personalities, in essence exaggerated versions of themselves, as described by Bill Oddie: "Tim is the respectable front-man, representative of the Establishment, Graeme plays the backroom boy who produces all the clever stuff and me, I'm the aggressive one." The Goodies runs for ten years on BBC, before moving to ITV for two seasons.

The Goodies cycle into our lives on their wobbly "trandem".

Education for All: The Open University

"The Open University changed my life and made me the person I am now."

Kathy Brooks, pioneer OU student, who sent her application form on the first day they could be received in 1971.

THE OPEN UNIVERSITY

The simple circle of a moon (O) in the dark sky of the U is a reminder of the night schools of the past, and how the Open University reinvents this tradition of adult learning.

In 1969 the Open University (OU) was created, based on Prime Minister Harold Wilson's earlier vision of "a university of the air". Then, starting on 3 January this year, students begin to study the first four OU courses, by a combination of written course materials, summer schools and – importantly – broadcasts on the BBC.

This unique partnership with the BBC transforms access to university education for many people. For the first time, tertiary students can learn at home, and can fit a degree around work and family commitments.

Open University programmes are frequently found at odd corners of the TV schedule – in the early mornings and late evenings, and are occasionally parodied by comedy shows such as *A Bit of Fry and Laurie* (1989) for their dated style and presentation (the notorious kipper ties) as programming is not regularly updated. But their impact is undeniable, and eventually the Open University will help over two million students to gain a degree and realize new ambitions.

Bob Hoskins as delivery man Alf who has problems reading and writing, in *On the Move*.

Other educational programmes
Meanwhile, School Radio has been impressively running programmes for schools since 1924. It is joined by School Television in 1957: *Living in the Commonwealth* is its first programme, followed by *Science Helps the Doctor*, *Spotlight on the Middle East* and *Young People at Work*. For primary school children, *Look and Read* (1967) transforms topic work, reading and writing. Later, the Continuing Education department will launch the motivational adult literacy campaign *On the Move* (1975) starring Bob Hoskins in his first TV role.

The Two Ronnies

"And now, it's goodnight from me – and it's goodnight from him."

Ronnie Corbett and Ronnie Barker

Ronnie Corbett and Ronnie Barker
in their inimitable end-of-show pose.

Ronnie Barker and Ronnie Corbett are both already established stars in their own right, when, following an impressive impromptu turn at the BAFTA Awards, the BBC brings them together to form the Two Ronnies partnership. The first show airs on 10 April and is an immediate success, running for 12 series and reaching an audience of 17 million at its peak.

The show has an established format, opening and closing with news headlines, and including Ronnie Corbett's seated monologues, filmed serials, such as the adventures of private detectives Charley Farley and Piggy Malone, and spectacular comic musical finales (their country and western singers Big Jim Jehosophat and Fat-Belly Jones proving particularly popular!).

Material is provided by the leading comedy pens of Barry Cryer, David Renwick, Spike Mullins, David Nobbs, Peter Vincent, and assorted Pythons. Barker contributes sketches under the pseudonym "Gerald Wiley", ensuring they are selected on their own merits.

Comparison is often drawn between the Two Ronnies and Morecambe and Wise because both are male duos, but the comic dynamics are very different: the latter are straight man and funny man, whereas the former have more mutable comedic roles. Indeed, both Ronnies pursue careers outside the partnership, Ronnie Corbett in *Sorry!* (1981) and Ronnie Barker in *Open all Hours* (1973) and *Porridge* (1974).

The Two Ronnies ends in 1988, though the pair are briefly reunited in 2005 for *The Two Ronnies Sketchbook*. Ronnie Barker sadly dies in 2005, and Ronnie Corbett in 2016.

Two of the nation's favourite clowns.

147

Parkinson

"It's not so much about what you ask as what you don't ask."

Michael Parkinson on the art of the interview.

Jazz singer Marion Montgomery performs on the first *Parkinson* show.

Frequently named the chat-show king of British television, Michael Parkinson launches his famous show on 19 June 1971. The *Radio Times* that week promises "conversation, guests, good music and the occasional surprise".

Parkinson's first guest is the American jazz singer Marion Montgomery. And over the next 11 years, Parkinson – or Parky as he will become known – interviews the great and the good and the occasionally dangerous, from David Niven, Kenneth Williams, Muhammad Ali, Orson Wells and Lauren Bacall, to Rod Hull and Emu (the bird attacks him: "All the wonderful interviews in my show – yet I'll probably be remembered for that bloody bird!"). The show is revived in the late 1990s and runs for another six years.

Parkinson's easy, commonsensical manner and plain-speaking Yorkshire persona endear him to viewers. At a time when celebrities rarely divulge much about their private lives, he manages to eke out revelations largely without awkwardness. His show becomes the benchmark for other chat shows across the decades.

Parkinson interviews footballing legend George Best in 1975.

148

In this year

3 January
Open University
programming begins.

17 February
Elizabeth R

10 April
The Two Ronnies

17 June
After Labour loses the
1970 General Election,
documentary *Yesterday's Men*
upsets former Prime Minister
Harold Wilson and members
of his Cabinet.

21 September
Late-night music show *The
Old Grey Whistle Test* caters
for "serious" rock music fans.

2 October
The Generation Game

19 October
Prisoner of war drama
Colditz.

24 December
The first in the series
A Ghost Story for Christmas,
"The Stalls of Barchester",
based on a story by
M. R. James.

Just a few of Parkinson's star
interviewees *(from top)*: Muhammad
Ali, Dame Margot Fonteyn, Yoko Ono,
David Beckham, Dame Edith Evans,
Lee Remick, Gina Lollabrigida, Elton
John, Peter Fonda, John Lennon.

149

The Generation Game

"Didn't he do well?"

Bruce Forsyth's classic catchphrase, to egg on the show's contestants.

**Bruce Forsyth
(1928–2017)**

Bruce Forsyth's career spans seven decades, from song and dance man to star Saturday night TV host. He begins in variety, eventually taking over as host of ITV's *Sunday Night at the London Palladium* and other entertainment programmes. His BBC career is ignited by *The Generation Game*, which makes him the best game-show host in the business in the 1970s and 1990s. Later in his career he returns to the BBC to co-host the phenomenally successful *Strictly Come Dancing* (2004). He is knighted in 2011.

Game shows proliferated on American television in the 1950s, and eventually make their way to Britain when ITV launches *Take Your Pick* and *Double Your Money* in 1955. Both prove enormously popular as they offer cash prizes. The BBC is unhappy with financial incentives, but cracks the genre when it launches *The Generation Game* on 2 October this year, hosted by variety star Bruce Forsyth.

Based on a Dutch TV game format, the show pits four families against each other in a sequence of comic tasks, with the winner choosing prizes off a conveyor belt at the end of the programme. Classic items famously include "a toaster, a fondue set, a cuddly toy!"

Apparently, when Head of Light Entertainment, Bill Cotton first saw the show on Dutch television, he said, "This has got Bruce Forsyth written all over it." He is proven right, as Forsyth is a huge hit and hosts the show from 1971 to 1977 and again from 1990 to 1994. At various times, Larry Grayson, Jim Davidson and Graham Norton also front the show, which is a key anchor in the BBC's Saturday night schedule for years, attracting 25 million viewers at its peak.

The Generation Game is revived in 2018, hosted by Mel Giedroyc and Sue Perkins, but does not repeat its past success.

Bruce Forsyth's peerless skill at interacting with the family contestants as well as the viewers at home helps make *The Generation Game* a massive hit.

The Thinker or the Circus Strongman? Bruce's classic pose
often appears in silhouette at the beginning or end of his shows.

Newsround

"Decades ago this was the first news programme I ever watched."
"We used to love getting the latest news before our parents!"

Two *Newsround* viewers remember the show's impact.

John Craven
(1940–)

John Craven begins as a presenter on topical news programme *Search* (1971), before getting his big break on *Newsround*. He becomes the well-known face of factual children's television, then broadens his appeal via a regular slot on the popular *Multi-Coloured Swap Shop* (1976) and its 1982 successor *Saturday Superstore*. In 1989, he makes a move to the rural affairs strand *Countryfile*, where he will remain as lead presenter for three decades and more.

News for children launched back in April 1950 with Children's Newsreel – in the same style as the Television Newsreel for adults, albeit with a different selection of stories and simpler language. Looking rather outmoded, it came off the air in 1961. A decade later, *Newsround* launches with a more child-friendly proposition.

It is fronted by John Craven – whose name soon becomes part of the programme title. An experienced presenter, he adopts a more casual demeanour than is usual for news, sitting on the desk rather than behind it, and dressed in shirt and tie and later a jumper, rather than a suit.

After a somewhat shaky start, *Newsround* provides useful background information to help children understand the key news stories of the day, as well as launching investigative reports on issues relevant to its young audience, from school dinners and uniforms to pocket money and animals. Bulletins are monitored to remove upsetting and violent material.

At a time when there is no rolling news service, *Newsround* also breaks major news stories, including the shooting of Pope John Paul II on 13 May 1981 and the explosion of the Space Shuttle *Challenger* on January 28 1986.

Shanequa Paris, one of the *Newsround* presenters, 2020.

The Question of Ulster

"We may have been dull but not dangerous."

Labour peer Lord Caradon on taking part in *The Question of Ulster*.

Ludovic Kennedy presents
a range of solutions to
The Question of Ulster.

On 5 January, 1972, the BBC broadcasts a three-hour television special – *The Question of Ulster: An Enquiry Into the Future*.

It is an example of BBC programming – this decade and beyond – that explores Northern Irish politics and risks highly inflammatory situations with the British government, often based on a supposition of what *might* be broadcast (and might present the IRA in a favourable light) rather than the actual content. The Ulster Unionists and the Home Secretary, Reginald Maudling, refuse to take part, demanding that the programme be banned. "The Full United Kingdom is Now in Peril" shouts the headline of the *Belfast News Letter*.

Seven and a half million people watch the programme on the night, including nearly two thirds of the population of Northern Ireland. It is a "cool, at times laborious, examination of eight different solutions to the problem of Ulster," comments Richard Francis, BBC's Northern Ireland Controller.

Inevitably, the final reaction to the broadcast is an anticlimax. Telephone calls to the BBC in advance of transmission are 10–1 against the programme; they shift to 5–1 in favour post viewing.

Several critics of the programme and some right-wing politicians admit that a more considered approach, as well as direct involvement in the broadcast might have been a better course of action. However, the Northern Ireland Troubles – and the heated concern around its coverage on air – does not go away…

153

Quizzing the Nation: Mastermind and University Challenge

"Oh, do come on!"

University Challenge quizmaster Jeremy Paxman's regular chiding of a slow-to-answer team.

Icelandic-born presenter and journalist Magnus Magnusson, the first host of *Mastermind*.

Quizzes become extremely popular post-war, with the launch of *The Brains Trust* (1951) and *Animal, Vegetable, Mineral* (1952), along with *Top of the Form* (1953) for school-age children, which runs for more than 20 years.

But the surprise hit is *Mastermind* which begins on 11 September, billed by *Radio Times* as a "new and exciting brain game", presented by the then little-known Magnus Magnusson. Avuncular though Magnusson is, the show has a deliberately intimidating atmosphere – apparently, its original producer, Bill Wright, was inspired by his Gestapo interrogations during World War II to create the show's bare set, sonorous music ("Approaching Menace," by Neil Richardson) and spot-lit black leather interrogation chair. The latter becomes the show's visual icon. That apart, the show has a simple format: each candidate must face a round of questions on a topic of their choice, followed by a round of general knowledge questions.

Magnusson is the question master for 25 years until 1997. He is succeeded after a short hiatus by journalist John Humphrys in 2003, who in turn passes the interrogator's baton to Clive Myrie in 2021.

Your starter for ten

University Challenge, the most erudite of quizzes, launched on ITV back in 1962. Chaired by the cool-as-cucumber Bamber Gascoigne, it ran for 25 years. It is revived by the BBC in 1994 under the chairmanship of Jeremy Paxman, whose famous brusquerie gives the programme a new lease of life. It also finds a fresh audience, helped by a broader catchment of the universities represented beyond the earlier Oxbridge domination.

Jeremy Paxman quizzes a new generation of students from 1994.

The old anonymous *Mastermind* chair is replaced in 2003 by the classic Eames "soft pad" design from the late 1960s.

In this year

5 January
The Question of Ulster

8 January
The first episode (of four) of *Ways of Seeing*, John Berger's series probing traditional assumptions about art.

4 April
Newsround

11 April
The unscripted *I'm Sorry I Haven't a Clue*, "the antidote to panel games", hosted by Humphrey Lyttelton.

11 September
Mastermind

28 September
Costume drama *War and Peace* brings Tolstoy's epic novel to the small screen.

"I've started so I'll finish."

Magnus Magnusson's famous catchphrase for completing a question when the end-of-round buzzer sounds.

That's Life!

> "MPs who normally criticized programmes they had never seen... knew their voters watched *That's Life!*, so they did, too. When we called for reform on behalf of our viewers, parliamentarians acted to change laws."
>
> Esther Rantzen

Esther gets her clipboard out in *That's Life!* in 1983.

On 26 May *That's Life!* in all its magnificent eclectic glory takes to the screen. Created by producer and writer John Lloyd, it is presented and produced by Esther Rantzen, based on an earlier consumer affairs show called *Braden's Week* on which they both worked. There is initial criticism of the show's mix of the serious and the humorous – including eye-catching press cuttings, notoriously rude vegetables, funny poems and songs from the likes of Richard Stilgoe, Cyril Fletcher and Victoria Wood. But the programme proves enormously popular and runs until 1994, with audiences in excess of 22 million at its peak.

Rantzen is the lead on-air presence in a television world still very much dominated by men. Over the years her co-presenters will include Glyn Worsnip, Kieran Prendiville, Chris Serle, Bill Buckley, Gavin Campbell, Paul Heiney and Adrian Mills.

The show is also one of the first programmes to make stars of the public in regular street interviews. In particular 84-year-old Annie Mizen charms the nation with her outspoken comments, sampling champagne one week ("I'm not a drinker really"), followed by Beluga caviar the next ("Tastes like mouse droppings... not that I've ever eaten 'em!").

Social action

At its core, *That's Life!* functions as the consumer's champion, but it also develops a very strong line in social action campaigning, especially around child-protection issues. In 1984, the case of Ben Hardwicke highlights the need for more child organ transplants and leads directly to a reduction in waiting lists. In 1986, the national

156

helpline Childline is set up after the show's survey on child abuse reveals the state of the need, and leads to the BBC One series *Childwatch*. Beginning with 100 volunteers in a small office, the helpline is soon busy, attracting 8,000 calls every day. Eventually, it will grow to a network of centres around the UK as well as an online service, and the model will be copied in over 150 countries.

Esther Rantzen
(1940–)

Her first key role at the BBC in 1986 is as researcher/reporter for *Braden's Week*, leading to her producing and presenting the groundbreaking *That's Life!* The show makes Rantzen a household name. Other TV highlights include her talk show *Esther, Hearts of Gold*, and a landmark programme on palliative care. Out of her programming come two new and important charities: Childline (1986) for children and young people and Silver Line (2012) offering information and advice to older people. She is made a Dame in 2015.

That's new life! – Esther Rantzen, Kieron Prendiville, Cyril Fletcher and Glyn Worsnip.

TV Chefs: From Cookery to Family Fare

"I call her Delia, so totally has she taken on the nature of a family friend."

Broadcaster Joan Bakewell, 1980

Cookery features on the BBC from its early days, with Marguerite Patten explaining how to make the best of basic ingredients during World War II, to Philip Harben (the first official "television chef") adding a dash of continental style from 1946, when he instructs viewers how to make an everyday lobster vol-au-vent (the opener dish on his *Cookery* show, which runs from 1946 to 1951)!

Fanny Cradock is the next big thing in the TV kitchen. With her penchant for glamour and assisted by her largely silent husband Johnnie, she arrives on our screens in *Kitchen Magic* (1955), rattling through a prodigious number of shows before her insensitive remarks to an amateur chef on *The Big Time* (1971) finish her TV career.

"The Delia Effect"

In complete contrast to Cradock is Delia Smith, who first appears with her direct and sensible demeanour in *Family Fare* on 29 September. Her most famous series, *Delia's Smith's Cookery Course*, follows five years later. Smith is in complete contrast to the continental aspirations

Delia in action for *Delia Smith Cookery Course*.

In this year

4 January
Set in rural Yorkshire, Roy Clarke's gentle sitcom *The Last of the Summer Wine* focuses on a trio of old men looking for fun.

5 February
The Wombles stop-motion animated series.

14 March
There's campy fun aplenty in Jeremy Lloyd and David Croft's department store sitcom *Are You Being Served?*

2 April
Landmark documentary series *Open Door* allows marginalized groups free access to television.

5 May
Renowned scientist Jacob Bronowski's *Ascent of Man* series explores humanity's scientific achievements.

26 May
That's Life!

23 July:
The annual *Radio 1 Roadshow* takes DJs and pop acts out of London to locations throughout the UK

29 September
Family Fare

of past cookery shows – rather her approach is clear and methodical, issuing as it does from the BBC's Continuing Education department.

Further series/book tie-ins appear in the 80s and 90s: *One is Fun* (1986) *Delia Smith's Summer Collection* (1993) and *Delia's How to Cook* (1998).

Delia's impact on UK taste and consumption is enormous. Supermarket shelves are often emptied in hours when she cites a particular ingredient in her recipes, most famously in 1995 when her series *Delia's Winter Collection* features cranberries; the previously little-used berry is bought up overnight.

Chefs such as Keith Floyd, Rick Stein, The Hairy Bikers, Ken Hom, Nigella Lawson, Madhur Jaffrey and many more ensure that cookery programmes remain a crucial part of BBC TV scheduling.

During the 1990s, cookery shows embrace the fashion for reality TV by introducing a competitive element, sometimes including celebrity contestants, in shows such as *MasterChef* (1990), its various spin-offs and *The Great British Bake Off* (2010).

Inspiring us in the kitchen *(clockwise from top)*: Keith Floyd, The Hairy Bikers, Ken Hom, Nigella Lawson, Rick Stein and Madhur Jaffrey.

The Family

> "I wanted to make a film about the kind of people who
> never got on to television."
>
> Paul Watson on *The Family*.

The first "fly-on-the-wall" documentary ever seen on British television arrives on 3 April. Its impact is huge, as a fascinated and shocked nation watch the goings-on of the working-class Wilkins family in Reading.

The family is made up of mother Margaret and father Terry (both 39), with their four children Marion, Gary, Heather and Christopher, along with Gary's wife Karen, baby Scott and Marion's fiancé Tom. They agree to be filmed for 18 hours a day over a three-month period in their small flat above a greengrocer's shop. "No TV family ever has dirty pots and pans," says Margaret in episode one, and the show is bold in showing life as it is.

The Wilkins family in the front row get ready for the real-life wedding of Marion and Tom on 25 May, 1974.

However, the Wilkinses are unhappy when they first see themselves on screen, claiming the programme has been unfairly edited. Audiences are divided, too, and there are calls for it to be banned. Nevertheless, eight million watch Marion and Tom get married in what is called "The television wedding of the year". And the fly-on-the-wall documentary is here to stay – *Driving School* (1997) and *The Cruise* (1998) are just two examples to come, along with *The Office* (2001), the perfect spoof of the genre.

Learner driver Maureen Rees at the wheel with long-suffering husband Dave in *Driving School*.

Innovation: Ceefax

Ceefax becomes a surprise hit, paving the way for the Internet.

Television is changing – and the notion of it as an interactive text service, something akin to later online services – comes nearer to fruition when Ceefax launches on 23 September, enabling the viewer to literally "see facts". The first teletext service in the world, it is a minority interest at the beginning, but receives a great boost once gaps in the television schedule begin to be filled with a selection of pages accompanied by music. Eventually, Ceefax will have 22 million weekly users and develop a cult following all of its own.

Allison: "Don't tell me you're still on MySpace?"
Hal: "We're more Ceefax people."
Ceefax enters the cult memory in *Being Human*, series 4

National Storytelling:
Pobol y Cwm and River City

"Bore da, Maggie Mathias." ("Good morning, Maggie Mathias.")

The first lines spoken in *Pobol y Cwm*.

National and local stories, enshrined in soap opera, start to be told on BBC TV from the mid 1970s.

The longest-running of these is BBC Wales' *Pobol y Cwm*, created by John Hefin and Gwenlyn Parry, and launched on 16 October 1974. The show focuses on the people who live and work in the fictional village of Cwmderi, and is an immediate hit.

It also launches at a time when Welsh-language programmes are continually under threat, so its popularity soon makes it an important factor in the battle to establish the Welsh language channel, S4C.

Cassie and Teg Morris
in the Deri Arms.

Merry Christmas from the cast of
Pobol y Cwm in 1990.

Trouble for Alex Murdoch
(Jordan Young) in 2019.

River City

Compared to the long-running *Pobol y Cwm*, *River City* is a relative newcomer. Created by Stephen Greenhorn and launched in 2002, it is Scotland's only current homegrown soap and its longest-running drama serial.

The show is set in the fictional Glasgow district of Shieldinch on the banks of the Clyde where working class meets upper crust. This is contemporary Scotland, true to Glasgow's industrial roots while reflecting the city's considerable cultural diversity.

River City has a faithful following and regularly pulls in 500,000 viewers.

"Gut-wrenchingly brilliant!"

A fan's comment on one of *River City*'s 2017 dramatic highlights.

The cast of *River City* on location in 2005.

The Good Life

"I suppose we must be rather a blot on the avenue's escutcheon."

Barbara (Felicity Kendal) to Margo (Penelope Keith).

The Goods get to grips with the realities of the green life.

Friends and neighbours: Tom and Barbara Good (Richard Briers and Felicity Kendal) and Jerry and Margo Leadbetter (Paul Eddington and Penelope Keith), 1977.

As the 1970s evolve, so does the idea of "going green", the desire to explore alternative, more sustainable ways of living. The resultant clash of cultures is nowhere better expressed than in John Esmonde and Bob Larbey's hugely popular sitcom, *The Good Life*.

First screened on 4 April, it pits going-green, unworldly couple Tom and Barbara Good (Richard Briers and Felicity Kendal) against their uptight, prosperous middle-class next-door neighbours, Margo and Jerry Leadbetter (Penelope Keith and Paul Eddington), alternately appalled and bemused by the Goods' attempts to live off the land in suburban Surbiton.

The show makes stars of all its lead players who go on to further TV and theatre glory, and manages to benignly poke fun at both middle-class values and experimental environmentalists. The series runs for 30 episodes but repeats ensure it is never off our screens.

Fawlty Towers

"Well may I ask what you expected to see out of a Torquay hotel bedroom window? Sydney Opera House perhaps? The Hanging Gardens of Babylon? Herds of wildebeest sweeping majestically across the plain..."

Basil's scathing rejoinder to a hotel guest complaining about the lack of a view.

"Enjoy your stay!" – a welcome to Fawlty Towers from Basil.

Meanwhile, somewhere near Torquay on the South Coast of England is the worst hotel in the world, Fawlty Towers. Co-written by John Cleese and his wife Connie Booth, *Fawlty Towers* is Cleese's first comedy after *Monty Python's Flying Circus* (1969) and debuts on 19 September this year.

Each episode is, in effect, an elaborately plotted farce, focusing on Cleese's deranged hotelier Basil as he attempts to hide his incompetence from his domineering wife Sybil (Prunella Scales). Loyal chambermaid Polly (Connie Booth) and linguistically challenged Spanish waiter Manuel (Andrew Sachs) make up the comic quartet. The character of Fawlty is based on a real hotelier Cleese encountered while filming in Torbay. Only 12 half hour episodes are ever made, but *Fawlty Towers* proves to have an enduring appeal. In 2000, it will be voted the best British television programme of all time in a BFI poll.

Reception committee: Sybil (Prunella Scales), Basil (John Cleese), Polly (Connie Booth) and, of course, Manuel (Andrew Sachs), 1979.

The John Peel Show

"I just want to hear something I haven't heard before."

John Peel

**John Peel
(1939–2004)**

Born John Robert Parker Ravenscroft, he reinvents himself as DJ John Peel at pirate station Radio London in 1967. He joins the newly established Radio 1 the same year, fronting *Top Gear* and *Night Ride* and, in time, becoming one of the most influential DJs ever. From 1998, he also hosts award-winning Radio 4 talk show *Home Truths*, in which ordinary members of the public get the chance to tell their extraordinary stories. Following his death, there are reports of Peel's sexual misconduct with a minor, previously acknowledged by Peel himself in an autobiography. Peel's son, Tom Ravenscroft, is a regular DJ on the BBC's alternative music station Radio 6.

From September 1975, John Peel begins his daily late-night show on Radio 1, which he will front for over 30 years. Produced until 1991 by John Walters, it changes the face of British pop music, promoting acts such as The Fall, The Smiths, and Pulp, introducing the nation to drum 'n' bass, hip hop, and, most famously, to punk rock in the 1970s.

Peel's reputation as an important DJ breaking unsigned acts into the mainstream is such that aspiring bands and musicians send him records, cassette tapes, and CDs hoping, not always in vain, for airplay and a supportive word from the droll, self-effacing Peel. Returning home from a three-week holiday at the end of 1986, Peel finds 173 LPs, 91 12-inch singles and 179 7-inch singles waiting for him.

Peel's show also includes the Peel Sessions: four or so songs recorded by a guest artist live in the studio, often providing the first national coverage to bands that will later achieve great fame. As a tribute to his extraordinary impact on British music, in 2012 the BBC names a wing of New Broadcasting House after him.

John Peel uncovers the wonders of vinyl collections around the UK. The *Evening Standard* will sum up the feelings of music fans all over the world when, following his death on 25 October 2004, it comments: "The day the music died."

Arena

> "One of the greatest television shows ever put together
> and sustained for 40 years."
>
> David Thomson, film critic and historian.

12 October
Comedy series *On the Move*,
written by Barry Took and
starring Bob Hoskins as a
removal man with reading
problems, focuses on adult
illiteracy in modern society.

Launched by Humphrey Burton, Head of Arts and Music, on 1 October, *Arena* opens with a conversation between celebrated actor Laurence Olivier and critic Kenneth Tynan. Its mission: to extend arts coverage to the mainstream. Its BBC antecedents are *Monitor* (1958) and *Omnibus* (1967), both of which have led the way in committed and wide-ranging exploration of the arts.

However, *Arena* will go even further in its experimentation and variety. Across the years, over 600 *Arena* shows will feature everything from stand-alone feature films, to cultural essays, to biographical portraits. It will showcase some of the most important filmmakers of the century and a dazzling array of topics, from punk to Princess Diana; George Eliot to Cindy Sherman; Elvis Presley and food to Nelson Mandela on Robben Island; the anarchic creativity of Spitting Image to the longevity of Desert Island Discs.

Humphrey Burton is succeeded by Leslie Megahey, Alan Yentob, Nigel Finch and Anthony Wall, all of whom curate this versatile and eclectic strand, characterized by the bobbing bottle of its opening titles with its message of constant surprise and revelation.

The "message in a bottle" opening titles of *Arena*, accompanied by Brian Eno's haunting "Another Green World".

I, Claudius

"I grew up on *Claudius* [from age 11]... Now I'm 54 and watched the whole series – again – in three days. Irresistible!"

TV viewer reflecting on *I, Claudius* in 2019.

Derek Jacobi as the ageing Emperor Claudius, determined to set the record straight about his family's past misdeeds.

Another burst of historical costume drama, this time adapted from the pen of Robert Graves. Jack Pullman transforms the novels *I, Claudius* and *Claudius the God*, into a 12-part series that begins on 20 September and has the nation gripped with Rome fever.

This story of decadent imperial life with its mix of political intrigue, sex, and violence proves a critical hit, and turns the series' lead actor, Derek Jacobi, into a star. Lengthy hours in make-up are required to transform the late-thirties Jacobi into the aged Emperor Claudius: "First they made a cast of my face, covering my head in plaster of Paris, with two straws stuck up my nose to breathe through for 40 minutes until it set. It was like being buried alive", he recalled.

The fine supporting cast features memorable turns from Sian Phillips as Claudius' scheming mother Livia, George Baker as Tiberius, Brian Blessed as Emperor Augustus, Christopher Biggins as Nero, Patrick Stewart as Sejanus, Sheila White as Messalina and John Hurt as Caligula. Both Jacobi and Phillips win BAFTAs for their roles.

I, Claudius proves that historical drama does not have to be boring, and the BBC will return later to classical themes in *Rome* (2005) as well as exploring documentary strands led by classicist Mary Beard.

Sian Phillips as the ruthless Livia, who is a match for Claudius:
"They won't allow me in [the Senate] because I am a woman, and they won't allow you in because you're a fool."

The Fall and Rise of Reginald Perrin

"He spits, pops his eyes, flares his nostrils… I still haven't worked out why he is funny, but he is."

Clive James on Leonard Rossiter as Reggie Perrin, *Observer*

In this year

23 March
Roy Clarke sitcom *Open All Hours* stars Ronnie Barker as Arkwright, a crotchety grocer with a speech impediment (*below*), and his errand-boy Granville (David Jason).

8 September:
The Rise and Fall of Reginald Perrin.

20 September
I, Claudius

2 October:
Magazine show *Multi-Coloured Swap Shop* livens up Saturday morning TV for children.

30 October
The Shirley Bassey Show airs.

Leonard Rossiter as Reginald Perrin, flanked by a baffled Sally-Jane Spencer and Pauline Yates, 1977.

Meanwhile, back in suburbia… Britain is in 1970s meltdown and one man is gripped by the dreary realities of commuter living in David Nobbs' bleakly comic sitcom. Leonard Rossiter is Reginald Perrin, who fakes his suicide in the opening titles, then, in the course of the series, progresses towards a nervous breakdown, finally ending up as a happiness counsellor.

Rossiter is surrounded by a cast of great British acting talent, from Geoffrey Palmer as brother-in-law Jimmy, always in the middle of a "cock-up on the catering front", Tim Preece as a grim-faced estate agent and John Barron as his platitudinous boss Charles Jefferson or CJ, constantly mixing his metaphors to Pauline Yates as his long-suffering wife Elizabeth.

The show is unusual in having a narrative, transforming the standard sitcom vehicle. It runs for three series until 1979, plus a Christmas Special in 1982 and a follow-up series *The Legacy of Reginald Perrin* in 1996.

The News Quiz

*"Six-foot boa constrictor. Free to a good home.
Very friendly, good eater, likes children."*

Unintentionally humorous ads or headlines sent in by listeners are a favourite feature of the programme:
this from an advert in the pets section of the *Sheerness Times Guardian*.

In this year

10 April
Late-night television
documentary series
Everyman investigates
moral issues.

12 April
John Sullivan's sitcom
Citizen Smith stars
Robert Lindsay as rebel-
without-much-of-a-cause
Walter "Wolfie" Smith,
leader of the so-called
Tooting Popular Front.

The first edition of this highly popular and influential Radio 4 quiz is broadcast on 6 September , with chairman Barry Norman and guests Alan Coren, Richard Ingrams, Russell Davies and Clive James. The original idea for the show comes from the inspired John Lloyd who has a hand in numerous highly successful BBC radio and TV shows, including *You and Yours* (1970), *Quote… Unquote* (1976), *Not the Nine O'Clock News* (1979), *To the Manor Born* (1979) and *QI* (2003).

News Quiz presenters range from Barry Took and Simon Hoggart to Sandi Toksvig, Miles Jupp and Andy Zaltzman. The format, though, remains largely unchanged, with acerbic commentators on the day's news, along with a few normally serious radio newsreaders competing for largely made-up scores.

You and Yours and Money Box

With the rise in consumer activism in the 1970s, BBC Radio also plays a role – alongside television's driving force, *That's Life!* (1973). Beginning on 4 October 1970 *You and Yours* bills itself as "The Citizen's Advice Bureau of the Air". It promises to "tackle topics of direct concern to you", and the first edition covers all aspects of home ownership, from money and rights to health and welfare. Over the years, the programme broadens its appeal, tackling wide-ranging consumer issues, and actively seeking interaction with its listeners. Presenters include Liz Barclay, Patti Coldwell, Sue Cook, Derek

Liz Barclay, one of the regular
presenter voices on *You and Yours*.

Cooper, Paul Heiney, John Howard, Jenni Mills, and John Waite.

Then, on 2 October of this year, along comes *Money Box*, to explore in-depth aspects of personal finance, and hold to account the increasing number of companies and providers involved, as well as passing on useful advice to those trying to make the most of their money. The show also expands to take on a regular, mid-week *Money Box Live* show.

Where to go for sage financial advice? Paul Lewis of *Money Box* has the answers, 2005.

"A champion of the underdog... a fearsome campaigner... a long-haired terrier biting the ankles of the financial establishment."

Money Box presenter Paul Lewis (from 2000) praised by *Saga Magazine*

In Touch and Does He Take Sugar?

Peter White, the BBC's first Disability Affairs Correspondent, 1995.

Radio also reaches out across the decades to a wide community of listeners, often not catered for elsewhere. The groundbreaking *In Touch* launches in 1961 specifically for blind and partially sighted listeners, presented by David Scott Blackhall. He is followed in the presenter's chair by Peter White who will go on to become the BBC's first Disability Affairs Correspondent in 1995. White will later be the first blind person to produce reports for BBC television news.

Back on radio, *Does He Take Sugar?* hits the airwaves this year, the show's title reflecting the ways many people with disabilities are treated socially – as unable to speak for themselves. Created by producer Thena Heshel, the series creates one of the first authentic platforms for people with disabilities' voices, views and stories.

Grange Hill

"It's about school – something everyone has to face in one shape or form. I always thought that television could do more for children than tame things like "Robin Hood", or endless period dramas. It also needed to do more for the working-class comprehensive school kids."

Phil Redmond

In this year

8 January
All Creatures Great and Small stars Robert Hardy (below) and Christopeher Timothy as vets in the Yorkshire Dales during the 1930s.

7 February
The search is on to discover *The Young Musician of the Year.*

8 February
Grange Hill

7 March
Pennies from Heaven

8 March
Douglas Adams' sci-fi comedy *The Hitchhiker's Guide to the Galaxy* airs.

3 December
The BBC Shakespeare project to televise every Shakespeare play begins with *Romeo and Juliet.*

Created by Phil Redmond and set in a comprehensive school in North London, *Grange Hill* breaks the mould of children's school-based drama to date with a show that is realistic, at times graphic, and focused on working-class kids.

Grange Hill's camerawork is usually at child-level, its characters are real and funny and memorable, and its storylines increasingly challenging, from bullying and shoplifting to pregnancy and drug addiction. One of its most significant stories features pupil Zammo's battle with heroin addiction, not the sort of content usually aired at tea time.

A spin-off series, *Tucker's Luck* (1983) follows lead character Tucker and two of his old Grange Hill friends as they adapt to life after school. Later, many of *Grange Hill's* young actors, including Todd Carty, Susan Tully, Letitia Dean, Michelle Gayle and Sean Maguire, will graduate to *EastEnders*, with Susan Tully also moving into TV directing.

In January 2022, Sir Phil Redmond (knighted in 2021) will announce that he hopes to create a new *Grange Hill* film in 2023, featuring some of the original cast in parental roles.

Todd Carty as Tucker in 1980, one of *Grange Hill's* most memorable characters.

Pennies from Heaven

"A writer helps you show things you knew but didn't know you knew."

Dennis Potter

Dennis Potter
(1935–1994)

Dennis Potter deliberately sets out to make television as powerful an artistic vehicle as theatre and film. He makes his BBC debut writing for *The Wednesday Play* and *Play for Today*, with single dramas such as *Blue Remembered Hills* (1979), which casts adult actors as children, and *Brimstone and Treacle* (1976, initially banned by the BBC and not transmitted for 11 years). This is followed by his huge hits, *Pennies from Heaven* and, in 1986, *The Singing Detective*, which explicitly draws on Potter's debilitating psoriasis condition. Both works use non-naturalistic devices to reveal the characters' inner worlds. Potter's later works include the serials *Blackeyes* (1989) and *Lipstick on Your Collar* (1993), before his untimely death in 1994.

Pennies from Heaven is the television drama that really makes writer Dennis Potter's name, following a string of single play successes. The six-part series begins on 7 March and tells the story of struggling music salesman Arthur Parker in the depressed world of 1930s Britain. What transforms his life are "pennies from heaven", romantic melodies from the records of the day – sung by the likes of Al Bowley, Lew Stone, Arthur Tracey – that lift him above the daily grind.

The innovative use of a soundtrack of contemporary songs, some of which the cast mime to on screen, becomes Potter's inspired trademark, and he will pursue it again in *The Singing Detective* (1986) and *Lipstick on Your Collar* (1993). But *Pennies from Heaven* is the first, and wins him a BAFTA for most original series and takes him to Hollywood for a film version starring Steve Martin. It also makes a star of lead actor Bob Hoskins, with great work from a cast including Cheryl Campbell, Gemma Craven and Hywel Bennett. Potter's work will be a huge influence on many TV writers that follow him.

Bob Hoskins and Cheryl Campbell share an idealized moment in *Pennies from Heaven*, "a musical play".

Life on Earth

"It's surely our responsibility to do everything within our power to create
a planet that provides a home not just for us, but for all life on Earth."

David Attenborough

"Your TV documentaries, new and old, have inspired our generation of young people to appreciate the beauty, complexity and fragility of the web of life that we are lucky to have on this planet." From a letter sent by Youth Climate Ambassadors to David Attenborough, 2020.

David Attenborough launches his first groundbreaking natural history landmark series, *Life on Earth*, on 16 January. It changes everything.

Produced by the BBC's Natural History Unit, three years in the making and filmed in over 30 countries, it is ambitious in the extreme – tracing the development of life from its earliest times up to the present moment. Many of the wonders of the world's wildlife are captured in breathtaking images, with Attenborough rarely centre of the picture, usually only present as a discreet and guiding voice.

The programme's most famous moment is when Attenborough meets a family of gorillas in the jungle, and engages in a surprising and intimate physical conversation with them: "There is more meaning and mutual understanding in exchanging a glance with a gorilla than any other animal I know" he observes. "We're so similar." Other TV "firsts" include the sight of the "living fossil" fish coelacanth, and the spectacular courtship displays of birds of paradise.

The success of the series confirms the BBC Natural History Unit as world leaders in natural history filmmaking. *Life on Earth* provides the template for several more Attenborough series and makes him one of broadcasting's most eloquent and popular communicators, across all generations.

David Attenborough's surprising and touching encounter with a family of gorillas in *Life on Earth*.

David Attenborough
(1926–)

He begins work at the BBC in 1952, eventually producing the natural history series that are his passion. Attenborough first appears on TV in *Zoo Quest* (1954), filling in for zoologist Jack Lester when the latter falls ill. He produces and commissions other shows, before becoming Controller of BBC Two (where he oversees the introduction of colour TV) and Director of Programmes. He returns to programme-making – his real passion – to develop his first landmark natural history series, *Life on Earth*. This is succeeded by *The Living Planet* (1984) and *The Trials of Life* (1990). Other projects follow, including *Life in the Freezer* (1993), a celebration of Antarctica, *The Private Life of Plants* (1995 and *Attenborough in Paradise* (1996). More significantly, his subsequent documentaries – *Planet Earth* (2006), *Blue Planet I* and *II* ((2001 and 2017), *The Green Planet* (2022) – are passionate platforms for Attenborough's message to save the world and fast from global warming, pollution and other destructive tendencies of humanity.

Attenborough gets ready to dive with sharks in the later series, *The Living Planet*, 1984.

175

The Antiques Roadshow

"We're lucky in this country that we have more antiques per square foot than anywhere in the world, so I think we have a way to go yet."

Ceramics expert David Battie on the show's likely longevity.

Fiona Bruce with eager *Antiques Roadshow* participants at Castle Ward, Northern Ireland, 2020.

Working on a BBC documentary about a London auction house touring the West Country, producer Robin Drake is fascinated to see the audience's reactions. And so *Antiques Roadshow* is born. The pilot episode, recorded in Hereford, is a great success and the format remains almost exactly the same across the decades.

The show begins on 18 February, broadcast from Newbury. Bruce Parker is the first presenter, to be followed by Angela Rippon, Hugh Scully, Michael Aspel and Fiona Bruce. The initial star of the show is Arthur Negus, already famous from the television antiques show *Going for a Song* (1965), which had finished just two years earlier. But it is the items brought along by the public and each owner's reactions – especially when an expert tells them what their item might make at auction – that makes *Antiques Roadshow* such compulsive viewing.

There are countless stories of discoveries and disappointments – from a prototype model of Antony Gormley's "The Angel of the North" sculpture valued at £1m to a glazed jug thought to be by Picasso which turns out instead to be a school art project from the 1970s (the latter featuring on the US version of the show).

Over its duration, *Antiques Roadshow* will inspire a wave of interest in antiques and collectables as well as a whole genre of programmes about discovering them, from *Bargain Hunt* (2000) and *Flog It!* (2002) to *Antiques Road Trip* (2010).

Question Time

"I think you ought to leave now."

David Dimbleby reaches the end of his patience with one insistent audience member.

"The hour to question the ideas and decisions of today" is how *Radio Times* describes the first *Question Time* on 25 September.

Fronted by arch-inquisitor Robin Day, the show is innovative in really giving ordinary people a chance to interrogate politicians and opinion formers of the day – very much in tune with the increasing openness of the times. The initial production team is also unusual (for the day) in being all-female, led by producer Barbara Maxwell.

Day is chair for ten years until 1989, succeeded by Peter Sissons (1989-93) and David Dimblebly who will do the job for 25 years, until Fiona Bruce takes over as its first female chair. On Dimbleby's last show, he thanks the audiences for exercizing "what is a really important democratic right, putting questions to the panel and to argue with each other".

The audience is vetted and the guests on the panel chosen to exhibit a range of political views. *Question Time* becomes a crucible for lively debate, and is occasionally the focus of controversy, as when there are protests about the appearance of Nick Griffin of the British National Party on the show in 2009.

Long-time *Question Time* chair
David Dimbleby invites a comment
from the audience, 2004.

FEED T

JULY 13th 1985 a

LIVE AID

The 1980s Conflict On and Off Air

War at home and abroad, and the global horror of starvation. The BBC is embattled, too, clashing with a determined and tenacious Prime Minister. But human kindness finds a new broadcast voice, in Live Aid and a yellow bear.

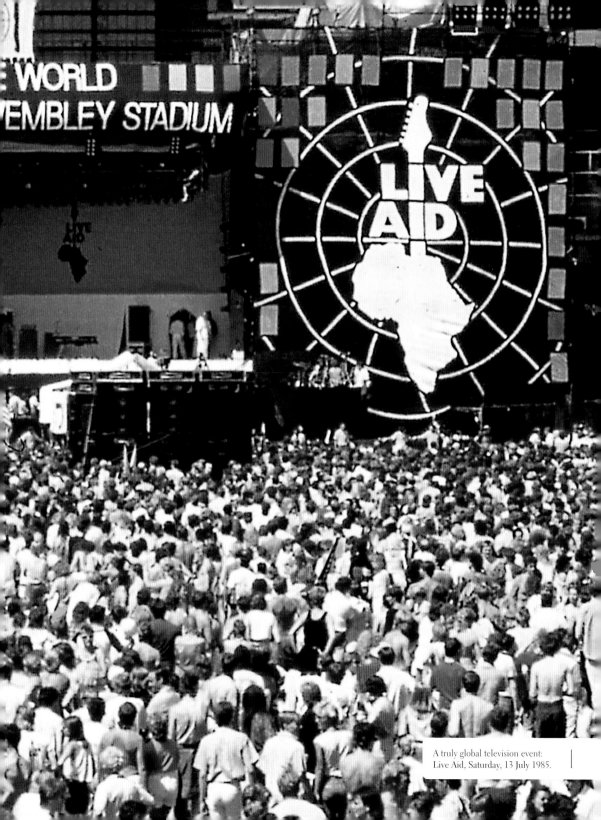

A truly global television event:
Live Aid, Saturday, 13 July 1985.

Newsnight

"I am always asking myself, why is this lying bastard lying to me?"

Newsnight's Jeremy Paxman opines on politics and the art of the political interview.

The very first *Newsnight* set.

The first edition of *Newsnight*, the BBC's flagship news-review programme, is broadcast on 30 January. Its aim is simple: to provide clear, in-depth analysis of the day's news. But maybe not so simple to deliver, as its content cuts across the News division and the Current Affairs division, which up to then function separately within the BBC. However, the show's founding editor George Carey is determined to pull it off, commenting later: "The opportunity to make a name for ourselves was there for the taking if we could just get our act together."

Experienced news presenter Peter Snow anchors the first show, with Fran Morrison presenting the news and weather and David Davis the sport. The main stories are Prime Minister Margaret Thatcher's attempts to get a rebate on British contributions to the EEC; curbs to trade union power; the Russian invasion of Afghanistan; and allegations of brutality by prison officers in Wormwood Scrubs.

Presented by...

Jeremy Paxman takes over in 1989, adding a new cut and thrust to the nightly proceedings.

Later, the presenter's chair is filled at various times by John Tusa, Charles Wheeler, Sarah Montague, Jeremy Vine, Kirsty Wark, among many others. Perhaps the programme's most famous incumbent is Jeremy Paxman (1989 to 2014), whose inquisitorial interviewing style redefines the programme. Paxman's most famous interview occurs on 13 May 1997 in which he asks Home Secretary Michael Howard the same question – whether or not he has overruled the director general of the prison service – 14 times in an attempt to elicit a straight answer. Some years afterwards, Paxman explains he was simply stringing the interview out because the next item was delayed...

Emily Maitlis wins RTS network presenter of 2020 for her interview with Prince Andrew.

Newsnight scores another *cause célèbre* interview in 2019 when presenter Emily Maitlis interviews Prince Andrew about his possible involvement in the Jeffrey Epstein sex-trafficking case. Maitlis recalls: "I realized within minutes of sitting down, it was going to be explosive. I don't think I'll ever forget it."

For several years, the programme suffers from erratic scheduling, but eventually, in 1988, settles at a regular time of 10.45pm. News-hungry viewers looking for in-depth analysis make a date not to miss it.

Yes Minister

"Clarification is not to clarify things. It is to put one's self in the clear."

Yes Minister's Sir Humphrey Appleby (Nigel Hawthorne) on the art of politics.

Derek Fowlds as Bernard Woolley, Nigel Hawthorne as Sir Humphrey and Paul Eddington as Jim Hacker MP in the 1984 Christmas edition of *Yes Minister*.

With its biggest fan Prime Minister Margaret Thatcher herself – she even appears in one episode demanding the abolition of economists, and quickly! – satirical sitcom *Yes Minister* has the seal of approval for the accuracy of its portrayal of civil servant shenanigans.

First broadcast on 25 February, written by Antony Jay and Jonathan Lynn, the show goes on to win six BAFTA awards, before transforming into the equally successful *Yes, Prime Minister*.

Boasting a sparkling cast of Paul Eddington as Jim Hacker MP, Nigel Hawthorne as devious Sir Humphrey Appleby, his Permanent Secretary, and Derek Fowlds as put-upon Private Secretary Bernard Woolley, *Yes Minister* transforms public perception of the machinery of government. Sir Humphrey's pronouncements become the stuff of "political spin" before the term is even invented.

Children in Need

"I think if they wanted to get me to leave Children in Need, they'd have to drag me off screaming. It's one thing that's so close to my heart, and I feel passionately about it."

Terry Wogan

The BBC's first charitable children's appeal began on Christmas Day 1927, raising £1,300. In 1955, television entered the arena with the Children's Hour appeal fronted by its first bear – Sooty – continuing until 1979.

Then on 21 November 1980, everything gets bigger and bolder with the first-ever *BBC Children in Need* telethon, using television in a more dynamic way to engage audiences. Made up of a series of short segments linking a whole evening's programming, its impact is immediate. It is helped, too, by its charismatic presenters: Terry Wogan (who will become inseparably linked to the charity's work), Sue Lawley and Esther Rantzen. The show raises £1 million. Something big is beginning.

Terry Wogan, Pudsey Bear and Sue Cook with a very large cheque, 1991.

Pudsey Bear, after his 2009 graphic makeover.

Introducing Pudsey

The famous Pudsey Bear makes his first appearance on *Children in Need* in 1985 as a brown, cuddly mascot. He is created by BBC graphic designer Joanna Lane, who names him after her hometown, midway between Bradford and Leeds.

Pudsey will have a number of makeovers across the decades, including a 2012 glamour re-costuming by celebrated designers, including Burberry, Victoria Beckham, Gucci, Prada and Alexander McQueen.

The charity will also become celebrated for its ingenious and different projects to raise money, from its Number One hit single "A Perfect Day" (1997) featuring the voice of the song's composer Lou Reed, plus Bono, David Bowie, Elton John, Boyzone and Tom Jones among others, to Matt Baker's Rickshaw challenges of 2011/12 and its Rocks for Terry concert (2016) celebrating the memory of Sir Terry Wogan.

In 2018, *Children in Need* will record an official total of £1 billion raised to support children and young people across the UK since 1980.

Dancing Newsqueens Susanna Reid, Fiona Bruce and Sophie Raworth perform Abba for *Children in Need* in 2008.

In this year

30 January
Newsnight

25 February
Yes Minister

21 September
Art critic Robert Hughes presents the documentary series *Shock of the New*.

21 November
The first major TV *Children in Need* appeal.

A Royal Wedding

"What a dream she looks, what a dream!"

Angela Rippon commentates on the arrival of Lady Diana Spencer in her spectacular Elizabeth and David Emanuel dress, which becomes a 1980s fashion-setter.

The wedding of Princess Elizabeth to Philip Mountbatten in 1947.

In broadcast terms, royal weddings grow in importance post war. The marriage of future queen Princess Elizabeth to Philip Mountbatten in 1947 is orchestrated pageantry: her dress is decorated with patriotic symbols, he wears his naval uniform adorned with medals earned in active service, and the event is broadcast live to an international radio audience. Her sister Princess Margaret's wedding to Anthony Armstrong-Jones in 1960 has a more contemporary glamour, and is the first royal wedding to be televised live.

The wedding of the century

However, the wedding of Prince Charles to Lady Diana Spencer on 29 July 1981 is the royal wedding of the century, and an occasion of television spectacle. It also marks a high point in the popularity of the royal family, watched by a global television audience of 750 million in 74 countries. In Britain, where a public holiday is declared, 28.4 million watch on BBC and ITV – the majority on the BBC – while 600,000 well-wishers line the streets of London.

It is the first royal wedding to be hosted at St Paul's Cathedral – chosen for its larger capacity – instead of Westminster Abbey. Every aspect of the wedding: the lead-up; the revelation of the dress with its 25-foot (7.6-metre)-long train; the actual ceremony including Diana's confusion over the groom's long list of names; the famous balcony kiss, is captured on camera.

Angela Rippon and Peter Woods begin their commentary at 7.45 a.m., followed by Tom Fleming on the carriage processions, and the marriage service beginning at 11.00 a.m. Simultaneous coverage on

TV wedding: Princess Margaret weds Anthony Armstrong-Jones in 1960.

Radio commentators Susannah Simons and Terry Wogan, 1981.

Lady Diana Spencer's famous 25-foot (7.6-metre) train follows her down the steps of St Paul's Cathedral.

BBC Two provides live subtitles for hearing-impaired viewers, the first big outing for the Palantype system. Radio coverage is also extensive, with commentators along the processional route including Wynford Vaughan-Thomas (from World War II broadcasting fame), and Radio 2's Terry Wogan.

The Royal Wedding remains one of the BBC's most-watched programmes, later viewed with special poignancy when the marriage unravels and both Prince Charles and Princess Diana seek documentary coverage to tell their sides of the story.

The royal weddings that follow on the BBC – Prince Edward and Sophie Rhys-Jones, Prince Harry and Meghan Markle – are more intimate affairs. However, in 2011, the wedding of future king Prince William to Catherine Middleton attracts a huge global audience, estimated at two billion.

Only Fools and Horses

"They're yuppies. They don't speak proper English like what we do."

Del Boy (David Jason) revels in the class tensions of the 1980s

In this year

26 February
Jimmy Perry and David Croft's nostalgic holiday-camp sitcom *Hi-de-Hi!* airs.

8 March
Radio 4 dramatizes the first episode of J. R. R. Tolkien's *The Lord of the Rings*.

29 July
The wedding of Prince Charles and Lady Diana Spencer.

27 August
Moira Stuart is the first Black woman to read the TV news.

8 September
Only Fools and Horses

11 October
See Hear, a TV programme for the deaf or hard of hearing community.

"Why do only fools and horses work?" questions this show's theme tune. This ever-popular sitcom follows the ill-fated adventures of wheeler-dealer Del Boy (David Jason) and Rodney (Nicholas Lyndhurst) Trotter looking to get rich quick in Peckham, South London. Schemes rarely work out, though Del Boy is a born optimist and "an enduring icon of the Thatcher era" (Darren Lee, BFI), while Rodney never quite manages to escape the lure of the possible.

The show is written by John Sullivan, fresh from his success with *Citizen Smith*, and hits our screens on 8 September. Sullivan creates a parade of immortal characters, including Grandad (Leonard Pearce), later replaced by Uncle Albert (Buster Merryfield) following Pearce's death, Trigger the road sweeper (Roger Lloyd Pack), posh car dealer Boycie (John Challis) and loser Denzil (Paul Barber). The Trotter brothers' two girlfriends, Raquel (Tessa Peake-Jones) and Cassandra (Gwyneth Strong), provide the only modicum of good sense in the ill-assorted company.

Del Boy (David Jason), Rodney (Nicholas Lyndhurst) and Uncle Albert (Buster Merryfield) parking the trademark Robin Reliant van.

David Jason (1940–)

At 25, David John White starts work as a theatre actor, adopting the name David Jason after his favourite film, *Jason and the Argonauts*. In the late 1960s, he begins his collaboration with Ronnie Barker, appearing in *Hark at Barker* (1969) and *Porridge* (1974), and as Granville in *Open all Hours* (1976). This is followed by his star turn as Del Boy in *Only Fools and Horses*, which he gains by making fun of director Ray Butt's Cockney accent with a pitch-perfect imitation. Notable non-BBC performances are as Pop Larkin in *The Darling Buds of May* (1991), Detective Inspector Jack Frost in *A Touch of Frost* (1992), and the voice of cartoon character Danger Mouse. He is knighted in 2005.

"Because he's Mr Everybody. We can all relate to him. Like him, we all think we can become millionaires and that will be the solution to all our difficulties, but of course it never works out."

David Jason on Del Boy's universal appeal.

"This time next year…"

The show makes stars of its leads, and spawns catchphrases that define the era, such as Del Boy's "Lovely Jubbly!" or his apocryphal "This time next year we'll be millionaires!" Show-stopping scenes include the famous descending chandelier incident in an episode that also features the immortal line: "Asking a Trotter if he knows anything about chandeliers is like asking Mr Kipling if he knows anything about cakes."

The series ends in 1991, but regular repeats and a string of one-off Christmas specials keep it on our screens. Spin-off comedies also emerge: *The Green Green Grass* (2005), where Marlene and Boycie, fleeing from gangsters, attempt to start a new life in the country, and *Rock & Chips* (2010) a prequel to *Only Fools and Horses*. Polls regularly cite the show as the favourite British comedy series of all time.

A lovely jubbly case of cutlery! David Jason as Del Boy on his way to making that million.

The Falklands War

"I'm not allowed to say how many planes joined the raid, but I counted them all out and I counted them all back."

Hanrahan's famous Falklands commentary line, 1 May 1982.

Brian Hanrahan (1949–2010)

During 20 years as a BBC foreign affairs reporter, Hanrahan covers most major world events, including the fall of the Berlin Wall in 1989, the rise of Mikhail Gorbachev in the Soviet Union and the struggles of the Solidarity trade union in Poland. Later, as Diplomatic Editor, he will provide live studio analysis of the 9/11 terrorist attacks. He establishes a reputation as an authoritative commentator with an eloquent economy of words in the tradition of the best broadcast correspondents.

On 2 April this year, Argentina invades the Falkland Islands, a remote UK colony in the South Atlantic. The move leads to a brief but bitter war. Argentina's military junta under General Galtieri aims to bolster support at a time of economic crisis by reclaiming sovereignty of the islands, while Prime Minister Margaret Thatcher is resolute that the 1,800 Falklanders are "of British tradition and stock", and despatches a task force to reclaim the islands.

In the fighting that follows (ending on 14 June), 655 Argentine and 255 British servicemen lose their lives, as do three Falkland Islanders. Without doubt, the most famous journalistic moment of the Falklands War is Brian Hanrahan's memorable "I counted them all out..." comment. Hanrahan uses this wording to get round reporting restrictions and reassure viewers that all the British Harrier jump-jets have returned safely from the Port Stanley raid of 1 May 1982.

The Falklands: the British cemetery at San Carlos, the site of the British landings on the Islands.

Model journalism

The BBC World Service covers the conflict with immediacy and clarity – even though many of its Latin American members have Argentinian connections. However, on the Home Front, the BBC clashes with the government, resulting in an incendiary encounter between Mrs Thatcher and the BBC. *Newsnight* presenter Peter Snow is accused of treachery for calling the troops "the British" rather than "our troops" or "our boys". The press also blames a BBC report for a soldier's death; however, that briefing later proves to have come from a government minister.

Prime Minister Mrs Thatcher, a strong advocate for the Falklands War.

Boys from the Blackstuff

"Gizza job. Go on, gizza job. I can do that."

The hectoring catchphrase of Yosser Hughes (Bernard Hill) captures the desperation of this period.

Bernard Hill as Yosser, whose life unravels once he is made unemployed.

Beginning life in January as a single drama entitled *The Black Stuff*, writer Alan Bleasedale's hard-hitting black comedy becomes a series, *Boys from the Blackstuff*, in October. Set in Liverpool, against the harsh backdrop of unemployment of Thatcher's Britain, it tells the story of a group of tarmac layers seeking work.

Nowhere is this plight better illustrated than in Bernard Hill's BAFTA-winning portrayal of Yosser Hughes, who is reduced to a shell of a person as he loses his job, family and self-respect.

Like *Cathy Come Home* before it, *Boys from the Blackstuff* has a strong impact on British society, showing the power of television to define a political and social moment in time through the searingly personal.

Breakfast Time

"What's wrong with the radio in the morning...? It got us through the war."

Breakfast Time's Editor Ron Neil recalls the initial adverse reactions to the idea of the show.

Good morning from the *Breakfast Time* team *(clockwise from top left)*: Francis Wilson, Debbie Rix, David Icke, Nick Ross, Selina Scott and Frank Bough.

In this year

17 January
Breakfast Time

23 March
Virginia Woolf's "unfilmable" novel *To the Lighthouse* is dramatized by Hugh Stoddart.

15 June
Blackadder

9 October
Charlotte Brontë's novel *Jane Eyre* is serialized, starring Zelah Clarke and Timothy Dalton.

24 October:
Replacing *Nationwide*, comes early evening current affairs show *Sixty Minutes*.

6 November
The first episode of Jane Austen's *Mansfield Park* airs, starring Sylvestra Le Touzel as Fanny Price.

29 November
Alan Bennett's *An Englishman Abroad* stars Coral Browne as herself and Alan Bates as Russian spy Guy Burgess.

On 17 January at 6.30 a.m. *Breakfast Time* launches, and our morning routine is changed forever. In fact, this is Europe's first-ever regular morning television service, pipping TV-am's *Good Morning Britain* by two weeks.

Presented by Frank Bough, Selina Scott, Debbie Rix and Nick Ross, with lighter segments from Russell Grant (astrology), the Green Goddess Diana Moran (keep fit), and Michael Smith and Glynn Christian (cookery), it pleasantly surprises audiences and critics by its easy-going and relaxed atmosphere. They are expecting a television equivalent of Radio 4's *Today* programme, not large red leather sofas and casual jugs of coffee and orange juice.

After the broadcast, the BBC receives thousands of calls from well-wishers saying how much they enjoyed it, and the show soon settles down to positive TV ratings.

Blackadder

"I have a cunning plan."

Baldrick's fail-safe catchphrase in a crisis.

Edmund Blackadder (Rowan Atkinson) in the show's first series, set in medieval times.

Blackadder II (clockwise): Lord Percy (Tim McInnerny), Lord Melchett (Stephen Fry), Baldrick (Tony Robinson), Nurse (Patsy Byrne), Blackadder (Rowan Atkinson), Queenie (Miranda Richardson).

History gets a comic makeover when Blackadder slinks onto our screens on 15 June, personified by Rowan Atkinson who co-writes the series with Richard Curtis. Initially, the show faces over-spending problems and is very nearly cancelled after the first run, but reinvents itself for the second series, drafts in the talents of Ben Elton as co-writer, and the rest is… history.

With each of its four series, the show changes period, the ambitious, scheming Blackadder (Atkinson) and his dozy servant Baldrick (Tony Robinson) being the key constants. Other regulars include Tim McInnerny (a foppish Lord Percy and Captain Darling), Stephen Fry (Lord Melchett, the Duke of Wellington and General Melchett) and Hugh Laurie (The Prince Regent and Captain George), as well as Miranda Richardson as an outrageous Queen Elizabeth I.

The end of the final season, set in the World War I trenches, provides an affecting climax as, in a slow-motion sequence, the men go "over the top" to certain death, and comedy mutates into tragedy.

Blackadder the Third (clockwise): Blackadder (Rowan Atkinson), Prince Regent (Hugh Laurie), Mrs Miggins (Helen Atkinson-Wood), (Baldrick Tony Robinson).

Blackadder Goes Forth (clockwise): Captain Darling (Tim McInnerny), General Melchett (Stephen Fry), Lt St Barleigh (Hugh Laurie), Private Baldrick (Tony Robinson), Captain Blackadder (Rowan Atkinson).

Real Lives

Belfast during The Troubles (1968–88).

Described in *Radio Times* as "a new series of filmed documentaries about the way people live now", *Real Lives* seeks to explore the wider world in all its diversity, from the Liverpool drug squad and the plight of the homeless to the aftermath of the Bhopal disaster and hairdressing competitions. But the episode that comes to define the series and creates one of the BBC's most seismic confrontations with government is "At the Edge of the Union", which investigates the Northern Irish political divide. It does this via the presentation of two parallel lives: Sinn Féin's Martin McGuinness and the Democratic Unionist's Gregory Campbell. "What is shocking," as *The Guardian*'s Charlotte Higgins later comments, "is that each talks war and death amid utterly ordinary, and strikingly similar, domestic existences – romping around on the beach with the kids, pouring milk from china jugs into cups of tea."

The fallout

Hearing of the programme, the Home Secretary Leon Brittan writes to the BBC urging them to withdraw it. Mrs Thatcher famously adds that any such programme will give "the oxygen of publicity" to the IRA. The BBC internally is at an impasse, and as a result BBC and ITN journalists stage a one-day strike in protest at the BBC's independence being compromised. Only eventually will a slightly amended version of the film be shown, in October 1985.

However, longer term, the effects of the clash are highly significant. In 1986, Mrs Thatcher brings in a new BBC Chairman, Marmaduke Hussey with an agenda of change, and, shortly afterwards Hussey dismisses Alasdair Milne, the Director-General: the first Director-General ever to be dismissed.

Crimewatch

"Don't have nightmares! Do sleep well."

Presenter Nick Ross's accidental remark at the end of the first show becomes his sign-off catchphrase.

First broadcast on 7 June, *Crimewatch UK*'s direct appeal to viewers to help with unsolved crimes and influence the delivery of justice is radical. "If you see anything tonight that jogs your memory please call us," is the message from joint presenters Nick Ross and Sue Cook.

Police officers and production staff are on hand to field calls, and the public is updated on any leads in a short programme later the same evening. After initial scepticism, police forces realize the value of the show in putting information directly before a large audience, and *Crimewatch* (as the show becomes known) viewers eventually assist the police with their enquiries on nearly 5,000 cases.

Many major crimes feature on *Crimewatch* over the years, including the mysterious 1999 murder of Jill Dando, a popular presenter of the show at the time of her tragic death. The programme ceases in 2017, but *Crimewatch Live* still features on BBC Daytime.

Jill Dando and Nick Ross on the *Crimewatch* set, 1998.

Victoria Wood: As Seen on TV

"You've probably never met a British person who doesn't like Victoria Wood...."

Hannah Mackay, Elegy to Victoria Wood, BFI

Victoria Wood
(1953–2016)

Spotted on talent show *New Faces* in the mid-1970s, Wood has a musical "residency" on *That's Life!* (1973), followed by one-woman shows and BAFTA–winning *Victoria Wood: As Seen on TV*. In 1998, *Dinnerladies* begins its run of two comedy series, written in parallel with more dramatic works such as *Pat and Margaret* (1994), the story of estranged sisters. Further triumphs follow with *Housewife 49* (2006), based on the wartime diaries of Nella Last and the musical *That Day We Sang* (2011).

This is the show that really showcases the star comedic talents of Victoria Wood. She creates her own comic world, one that is acutely observed and northern-based; its hilarity often underwritten by darker, poignant notes.

As the title suggests, much of this show riffs off the tropes of television: spoof documentaries (including the surprisingly touching story of ill-prepared cross-Channel swimmer Chrissie); Northern kitchen-sink drama replete with whippet; daytime TV presenters; hilariously snooty continuity links; and, of course, the all-time favourite "Ballad of Barry and Freda", which pastiches Cole Porter's "Let's Do It" to delicious effect, especially in its last verse: "Not bleakly/Not meekly/Beat me on the bottom with a *Woman's Weekly*/Let's do it, Let's do it tonight…"

The show also introduces "Acorn Antiques", Wood's parody of low-budget, daytime/early evening soaps, featuring rickety sets, implausible scripts, panicked and wooden acting, and quirky camera angles. Starring her "company" of actors – long-time collaborator Julie Walters as comic creation Mrs Overall joined by Celia Imrie and Duncan Preston – it is such a hit that it transforms into a stage musical in 2005.

Anyone for "Acorn Antiques", featuring Julie Walters, Duncan Preston, Victoria Wood and Celia Imrie, 2017?

"Jogging is for people who aren't intelligent enough to watch television"– words of wisdom from Victoria Wood, 1992.

194

Wogan

**Terry Wogan
(1938–2016)**

His first regular BBC Radio show is *Midday Spin*, followed by a *Late Night Extra* slot in 1967 on the newly created Radio 1 – he commutes from Dublin! Then in 1972, Wogan begins his long-running relationship with Radio 2, hosting hugely popular morning show *Wake Up to Wogan* until 1984, and returning again from 1993 until 2009. It becomes the most listened to radio show in Europe. He shifts to television to become chat show host of *Wogan* (1985) for the next seven years. Wogan is also famous for his hilarious commentaries of the Eurovision Song Contest and his tireless support for the BBC's Children in Need charity, as well as presenter slots (starting in 1973) on *Come Dancing*, and comedy panel show *Blankety Blank* (1979). He is knighted in 2005.

"I'm glad there's been so much laughter in the audience tonight. But they're not laughing with you. They're laughing at you."

Terry Wogan to David Icke during his notorious interview in 1991.

Terry Wogan is already well-known on the BBC when his chat show *Wogan* launches on 18 February. It runs at 7 p.m., three evenings a week, as part of a revamped BBC One schedule. It aims to bring some of the relaxed charm of Wogan's radio show to the format, and to carry on the successful chat-show formula of *Parkinson* (1971). *Wogan* is an instant success, and the show runs for seven years, making Terry Wogan a ubiquitous fixture on BBC radio and TV networks.

Guests on the first show are Wendy Richard, about to star this year as Pauline Fowler in the launch of *EastEnders*, impressionist Rory Bremner and pop superstars Elton John and Tina Turner. Wogan features many stellar and surprising guests on his show, from Princess Anne, James Stewart and Kenneth Williams to trickier interviewees such as a noticeably "refreshed" George Best, David Icke (the sports presenter memorably claims to be "The Son of God" on the show) and a clearly extremely nervous Anne Bancroft.

The amazing career of Terry Wogan *(from left)*: *Blankety Blank*, 1979; with Bo Derek on *Wogan*, 1984; Eurovision with British winner Katrina, 1987.

EastEnders

"Having done the necessary research (by slowing down the theme tune so we can clearly count the beats), we can confirm that there are in fact nine "doof doofs", and anyone who says otherwise should be publicly shamed!"

Digital Spy on how many electronic drumbeats underline the cliffhanger at the end of each *EastEnders* episode.

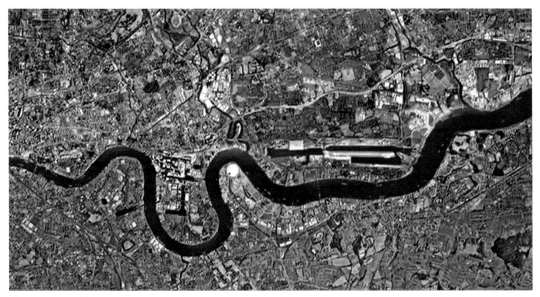

The famous map of the *EastEnders* opening titles, created by graphic designer Alan Jeapes and composed of more than 800 aerial photos.

The centre of the soap action: Albert Square in the fictional borough of Walford, E20.

Thirteen million viewers watch *EastEnders'* first episode, broadcast on 19 February. It begins with a (literal) bang as Arthur Fowler, Ali Osman and Den Watts break into Reg Cox's flat to find Reg near death. Only weeks later is Nick Cotton revealed as Reg's murderer.

Gossip, intrigue and scandal

EastEnders is created out of strategic necessity: to rival ITV's *Coronation Street* (going strong since 1960) and attract big audiences to BBC One. Its producer and writer are Julia Smith and Tony Holland, the latter drawing on his own East End childhood to create a drama set in a fictional London borough (Walford, E20), full of competing family networks (the Beales and the Fowlers) and with

Den and Angie (Leslie Grantham and Anita Dobson) in the Queen Vic pub in happier days, before their marriage hit the rocks.

strong matriarchal figures. Introducing the soap in the *Radio Times*, Holland writes that "gossip, intrigue and scandal are high on the list of daily events".

However, *EastEnders does* not shy away from difficult social issues, including HIV and AIDS, mental health, domestic violence, and drug misuse, as well as subjects such as education and multiculturalism. It quickly becomes a sounding board for the nation, is never out of the tabloids, and brings in huge audiences – famously a record 30.5 million viewers watch on Christmas Day 1986 as "Dirty" Den Watts serves divorce papers to his wife Angie.

Soon, the soap is a lynchpin of the BBC schedules: 'When *EastEnders* is going well, BBC One is going well," says BBC Executive Lorraine Hegessey.

Clever casting is important to the soap's success. Many faces are unfamiliar, but celebrity faces appear, too, including Wendy Richard – famous as the glamorous Miss Brahms in *Are You Being Served?* (1972) – playing against type as dowdy Pauline Fowler, or Barbara Windsor taking over the Queen Vic as Peggy Mitchell and returning authentically to her East End roots.

The *EastEnders* cast in 2003.

Live Aid

Bob Geldof, Janice Long, Adam Ant,
Elton John, Gary Kemp, Tony
Hadley and Midge Ure pose at a
press conference for Live Aid in
Wembley Stadium, 1985.

Bob Geldof wears his message.

One of the biggest rock concerts ever staged, Live Aid is the brainchild
of Bob Geldof, leader of Irish new wave band The Boomtown Rats.
Geldof is spurred to philanthropic action after seeing Michael Buerk's
BBC news report of the Ethiopian famine in October 1984. He
mobilizes the top singers of the day into a group called Band Aid and
launches the chart-topping charity single "Do They Know It's
Christmas?" He then goes on to bigger things with the creation of Live
Aid on 13 July…

Creative tactics and speed

As production manage Andy Zweck will later say of Geldof's creative
tactics: "He had to call Elton [John] and say 'Queen are in and
Bowie's in', and of course they weren't. Then he'd call Bowie and say
'Elton and Queen are in.' It was a game of bluff."

The show is put together incredibly quickly, with sparse facilities
and no time for pre-event technical rehearsals. On the day itself,
Wembley Stadium is packed and TV pictures are beamed to over

198

1.5 billion people in 160 countries, via BBC Television Centre. At the same time, a transatlantic concert continues in the US, at the John F. Kennedy Stadium in Philadelphia.

Geldof manages to book over 50 of the music industry's biggest names, all giving their services for free: Queen, David Bowie, Elton John, Bob Dylan, The Who, Paul McCartney, U2, Madonna, Duran Duran and Paul Simon, to name but a few. The opening act is Status Quo playing "Rocking All Over the World". Famously, Phil Collins zooms across the Atlantic on a Concorde supersonic airliner to perform in both concerts.

Overall, Live Aid raises more than £92 million. There is later dispute about how much money goes where, but Live Aid's impact is undeniable: it puts the Ethiopian famine at the forefront of the world's consciousness and conscience, transforms charitable fundraising and sets the model for high-profile music benefits to follow.

72,000 fill Wembley Stadium for the Live Aid Concert, and a further 100,000 in Philadelphia.

Casualty

"It is the realism of it which makes the show sustainable…
If we didn't make it real, it would become melodrama very quickly."

Casualty Producer Jonathan Young

In this year

4 April
The first *Comic Relief* fundraiser.

1 May
Carla Lane's sitcom *Bread* focuses on the misadventures of the working-class Boswell family in Liverpool.

6 September
Casualty

13 October
Selina Scott and Jeff Banks give fashion advice and news in *The Clothes Show*.

30 October
Esther Rantzen and Sue Cook present *Childwatch*, advising children on protecting themselves from cruelty, as well as whom to contact via free helpline Childline.

16 November
The first episode of Dennis Potter's controversial drama *The Singing Detective*.

Several medical dramas begin to emerge on British television post-war. ITV scores several major successes, including *Emergency Ward 10* (1957), while the BBC charms audiences with the nostalgic *Dr Finlay's Casebook* (1962), set in the Scottish Highlands during the 1920s. *Angels* (1975), which runs for 220 episodes until 1983, takes a more realistic approach in its portrayal of young nurses at the fictitious St Angela's Hospital in Battersea.

Comedy and heroics

Then, on 6 September, the longest-running primetime medical drama in the world, *Casualty*, hits our screens. The hard-edged action is set in the Accident and Emergency Department of Holby City Hospital, a fictionalized version of Bristol Hospital. The series is created by Jeremy Brock and Paul Unwin, who are inspired by the "comedy and heroics" of everyday life in the National Health Service.

"Keep calm and trust Charlie Fairhead!" – a classic World War II slogan, adapted to celebrate the impact of stalwart *Casualty* character Charlie Fairhead, played by Derek Thompson.

A harrowing moment in the daily routine of hospital life.

The first series introduces the regular members of the night shift, including nurses Charlie Fairhead (Derek Thompson) who will become a lynchpin character and a vital point of calmness and trust; Megan Roach (Brenda Fricker); Clive King (George Harris); and Lisa "Duffy" Duffin (Cathy Shipton). In the opening episode, they are on the front line, dealing with the aftermath of a chemical spill at the docks, along with awkward patients and a stolen bunch of flowers – setting the tone for *Casualty*'s wide range of major and minor human stories. From then on, a constant stream of engrossing patient stories provides the drama that makes a peak-time weekend hit. The show is also not afraid to deal with such issues (controversial at the time) as HIV and AIDS.

Casualty soon expands beyond its initial 15-week commission to run more or less all year round. It leads to spin-off drama *Holby City* in 1999, as well as daytime soaps such as *Doctors* (2000) and the more medically explicit, harrowing drama of Jed Mercurio's *Bodies* (2004).

The cast of *Holby City* assemble for the 1000th episode, 2019.

French and Saunders

"I think [French and Saunders] have been great for women and they are
of themselves just incredibly funny whether they are male or female."

Comedian Jo Brand

Following *Victoria Wood as Seen on TV* the next big female comedy TV hit is French and Saunders, debuting on 9 March. The duo come out of the alternative comedy circuit of the late 1970s to give a new female take on the comedy duo. Forget Morecambe and Wise or The Two Ronnies, this is a sharply observed, media-savvy sketch and song show, which plays on the friendship of the two women as well as their shape-shifting preparedness to try anything.

Spoofs and send-ups

The first episode is a deliberately shambolic affair, with support from parodic house band Raw Sex – Roland Rivron and Simon Brint – and The Hot Hoofers dance troupe. Guest star Alison Moyet is on to sing a song – but first has to witness French do her own version…

More faces of Dawn French and Jennifer Saunders in series six of their comedy sketch show, 2004.

Gentlemen prefer… French and Saunders, in their spoof of Marilyn Monroe and Rosalind Russell, 2017.

Later shows become celebrated for their increasingly elaborate sketches, with hilarious music spoofs of Elton John, Abba, Madonna, Bananarama and The Corrs (among many others)as well as ornate film parodies of everything from *Whatever Happened to Baby Jane?* to *Titanic* and *The Exorcist* (where French displays extreme demonic tendencies and Saunders in doctor-mode responds with: "I think we're looking at cystitis!").

Both Jennifer Saunders and Dawn French go on to become individual stars in the comedy universe. A sketch in the show's third series inspires Saunders to write and star in *Absolutely Fabulous* (1992), which will become one of the biggest comedy hits of the next decade. French also has huge success as Reverend Geraldine Granger in Richard Curtis' *The Vicar of Dibley* (1994), and later as a novelist.

Talking Heads

"A fully peopled, entirely credible, hysterically funny, heartbreaking world. Technically, each one was a masterclass in writing and acting."

Guardian TV critic Lucy Mangan

Writer Alan Bennett behind the scenes.

An old broadcasting adage goes that "talking heads" make bad television, but not when you have Alan Bennett writing the script! Featuring some of the best British acting talent – Thora Hird, Maggie Smith, Stephanie Cole, Julie Walters, Patricia Routledge and Bennett himself – these beautifully crafted television monologues reinvent the format, and paint unforgettable portraits of damaged lives and thwarted expectations.

A second, equally acclaimed series follows in 1998, and the original monologues are revived – plus two new commissions – with new actors in the pandemic year of 2020. Earlier BBC work by Bennett includes the play cycle *Objects of Affection* (1982), *An Englishman Abroad* (1983) and *A Question of Attribution* (1991).

Thora Hird speaks out in *A Cream Cracker Under the Settee*.

Tumbledown

"I felt very much in the firing line when the film came out."

Director Richard Eyre

In this year

19 April
Talking Heads

9 May
Produced by Janet
Street-Porter, *Def II*
aims to capture
a teenage audience.

31 May
Tumbledown

24 July
The British countryside and
rural issues are explored in
Country File (later
Countryfile).

30 October
Radio 4 series *All in the Mind*
explores the potential and
limits of the human brain.

13 November
The first episode of *The
Chronicles of Narnia*, based
on the C. S. Lewis novels.

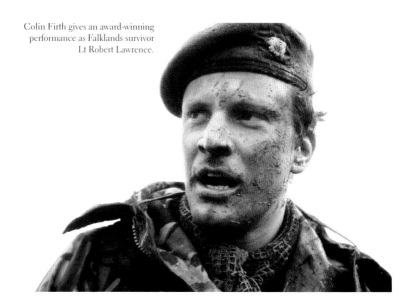

Colin Firth gives an award-winning
performance as Falklands survivor
Lt Robert Lawrence.

Towards the end of the troubled 1980s, Britain and the BBC reflect on
the Falklands War and its personal and political legacy. Within five
years, five dramas are screened, the most compelling of which is
Tumbledown, written by Charles Wood and directed by Richard Eyre.
Based on the experiences of Lt Robert Lawrence, who was seriously
injured during the war, the play features a searing lead performance
from Colin Firth. He portrays Lawrence in action, but also in the
aftermath of war as, struggling to adjust to life back home following
a paralyzing bullet wound, he is forgotten by the army.

Aired on 31 May, the drama attracts a surprising 10.5 million
viewers, helped in part by the extreme controversy generated around
it in advance. Its perceived anti-government, anti-war bias is even
questioned in Parliament. Despite such criticism, *Tumbledown* wins
acclaim, garnering a BAFTA for best play and an RTS award for Firth
as best actor.

A Bit of Fry and Laurie

"I was thin, practically without any visible talent but keen to use words
and an awkward frame to the best comic effect."

Stephen Fry

Stephen Fry and Hugh Laurie met at Cambridge University, and their clever, spoofy humour harks back to earlier Oxbridge comedy hits on the BBC such as *Not Only… But Also* (featuring Peter Cook and Dudley Moore, 1964), *Monty Python's Flying Circus* (1969), *The Goodies* (1970) and, following after them, *Armstrong and Miller* (1997) and *That Mitchell and Webb Look* (2006).

Structured loosely as a sketch show, *A Bit of Fry and Laurie* is variously eccentric, highbrow and downright silly. It frequently breaks the fourth wall, ricochets between characterization and real life, with piano playing (by Laurie) and riffing catchphrases. It makes stars of Fry and Laurie, and plays for four series.

The duo continue their witty banter into the 1990s, when they are cast in the title roles of ITV's *Jeeves and Wooster* (1990), before embarking on very different careers on different sides of the Atlantic. Laurie has huge success as a serious actor in the US medical drama *House* (2004), while Fry becomes a national treasure esteemed for his writings, openness about his own mental health, as well as his eclectic erudition as host of *QI* (2003).

"The reason we're not going to
do this sketch is that it contains a
great deal of sex and violence…"
Hugh Laurie and Stephen Fry
get eccentric and hilarious in
A Bit of Fry and Laurie.

Birds of a Feather

"People think they know me from somewhere, and then I open my mouth and they realize it's me from *Birds of a Feather*."

Pauline Quirke

Birds of a Feather has a long run – until 1998 – and attracts audiences as high as 20 million. It then shifts to ITV in 2014 for a further six years.

Meanwhile, a very different duo creating comedy are real-life friends Pauline Quirke and Linda Robson as sisters Sharon and Tracey, who are brought together when their husbands are sent to prison for armed robbery. Hard-up Sharon moves out of her council flat into Tracey's expensive home in Chigwell, Essex and so the comedy of opposites begins: rich and poor, fat and thin... It is often sparked by their outrageous neighbour Dorien Green (a star turn by Lesley Joseph), who effortlessly injects sexual innuendo into every conversation.

Written by Laurence Marks and Maurice Gran, the first episode of 16 October causes some complaints because of its sexual frankness, but the show is an immediate hit, its caustic, sparky dialogue capturing the new openness about materialism, money and sex.

Challenge Anneka

Unfailingly energetic problem-solver Anneka Rice in 1989.

And, somewhere in constant motion is jump-suited Anneka Rice, in her buggy or her helicopter off on her next challenge. Time to do good, and Anneka pulls off community project after project, building playgrounds, bridges, theatres and famously a Romanian orphanage – and persuading local businesses and volunteers to help her. Audiences of 11 million watch every Friday night for seven seasons till 1995.

"The affection for it is epic. So many people were touched by the series – people's lives were literally changed."

Anneka Rice

The 1990s
All Change

Everyone is live and kicking and behaving badly or absolutely
fabulously. Change is definitely in the air – your décor, your
cuisine... Oranges are not the only fruit, and even a Jane
Austen classic receives a makeover.

The *Changing Rooms* team ready for action *(from left)*: Laurence Llewelyn Bowen, Anna Ryder-Richardson, Andy Kane, Michael Jewitt, Linda Barker and Graham Wynne.

Oranges Are Not the Only Fruit

"A deeply moving and brilliantly acted depiction of a very personal story – a story that continues to change the lives of people who come across it today."

TV Heaven website

In this year

4 January
David Renwick's sitcom *One Foot in the Grave* features an irascible pensioner (Richard Wilson) and his long-suffering wife (Annette Crosbie) grappling with modern life, and spawns the catchphrase "I can't believe it!"

10 January
Oranges Are Not the Only Fruit

5 February
The time pips from Greenwich are heard for the last time.

2 July
Contestants enter the *MasterChef* kitchen.

27 August
Specializing in topical discussion and sports coverage, Radio 5 begins broadcasting.

Adapted by Jeanette Winterson from her semi-autobiographical novel, *Oranges Are Not the Only Fruit* bursts onto our screens on 10 January in a blaze of pre-publicity. It is one of the first high-profile dramas to feature a lesbian storyline – though the first lesbian kiss appeared much earlier on BBC in James Robson's 1974 drama *Girl*, starring a young Alison Steadman. But *Oranges Are Not the Only Fruit* foregrounds diverse sexuality much more boldly and positively, and paves the way for lesbian dramas *Tipping the Velvet* (2002), *Fingersmith* (2005) and *Gentleman Jack* (2019).

The series tells the story of Jess – an adopted child brought up in a strict Pentecostal family – who rejects her upbringing as she comes to realize who she is. Jess (Charlotte Coleman) battles her fanatical mother (Geraldine McEwan in a BAFTA-winning performance). The programme also wins the BAFTA for Best Drama.

Winterson's adaptation is directed by Beeban Kidron. Surprising camera angles regularly make the everyday world seem strange. Winterson writes of her desire to challenge "the virtues of the home, the power of the church and the supposed normality of heterosexuality", and *Oranges Are Not the Only Fruit* remains a compelling human story.

Jess (Charlotte Coleman) tries to escape the powerful presence of her mother (Geraldine McEwan).

210

Have I Got News for You

> "I've never been disappointed by politicians. I've never invested
> that much in them in the first place."
>
> Paul Merton

In this year

28 September
Have I Got News for You

3 October
David Attenborough's
documentary series *The
Trials of Life*, focuses on
crucial aspects of animal
behaviour..

8 November
Sketch show *Harry Enfield's
Televison Programme*
(renamed *Harry Enfield &
Chums*) stars Enfield, Paul
Whitehouse and Kathy Burke
in a variety of roles, including
the Old Gits, the Slobs and
DJs Smashie and Nicey.

TV satire has been in
short supply on the
BBC, since its brief
flourishing in the
1960s with *That Was
the Week That Was*
(1962) and *The Frost
Report* (1966),
followed by *Not the
Nine O'Clock News*
(1979), though
Spitting Image (1984)
is alive and well on
ITV. All that changes
with the arrival of
Have I Got News for You on 28 September.

The original *HIGNFY* line up of Ian Hislop,
Angus Deayton and Paul Merton.

Like many TV shows, it originates from radio – it is loosely based on
The News Quiz (1977). Its first chair is Angus Deayton and the two
teams are captained by *Private Eye* magazine editor Ian Hislop and
wry comedian Paul Merton, with support from Sandi Toksvig and Kate
Saunders. The ongoing success of the programme lies in its constant
rotation of guest panellists drawn from the varied world of
entertainment, journalism and politics, and frequently the butt of
many of the show's jokes. Deayton is replaced as chairman in 2002
after his private life becomes a tabloid news story, but the programme
continues to thrive with a spectacular mixture of guest presenters,
including PM-to-be-Boris Johnson, Bruce Forsyth, Jeremy Clarkson
and Charlotte Church, as well as prominent actors and comedians.

In 2001, *Have I Got News for You* wins the British Comedy
Academy Lifetime Achievement Award, the first time it is given to a
programme rather than an individual.

Comic Relief – "The Stonker"

"Comic Relief is like my fifth child – I'll never leave it and never not love it."

Comic Relief co-founder Richard Curtis

**Lenny Henry
(1958–)**

Henry comes to prominence as a finalist in ITV's *New Faces* talent show (1975), going on to present the kids' show *TISWAS* (1978) before moving into alternative comedy and meeting his future wife Dawn French. He has a string of comedy shows on the BBC, from *The Lenny Henry Show* (1984) to the sitcom *Chef!* (1993), making him one of the first mainstream Black comedians on UK TV. Serious acting roles follow, script writing and drama development, as well as the co-funding of *Comic Relief*, plus a key role in promoting Black voices in the media. He is knighted in 2015.

Having taken a couple of years off from the previous two Red Nose Days, *Comic Relief* returns this year with "The Stonker", the biggest and boldest fund-raising comedy night, so named because of the Hale and Pace official *Comic Relief* song "The Stonk" is performed on it – with Pink Floyd's David Gilmour on guitar and Mr Bean (Rowan Atkinson) on drums.

The show runs for 480 minutes, with hilarious turns from Ben Elton, Victoria Wood, French and Saunders, Ronnie Corbett, and many other stars.

What about a red nose?

The first night of *Comic Relief* took place back in 5 February 1988, hosted by Lenny Henry, Griff Rhys Jones and Jonathan Ross. The charity was founded a few years earlier by Richard Curtis, Jane Tewson and comedy friends. It built on the success of Band Aid and Live Aid, but used comedy rather than music to fundraise for people living in poverty in Africa and around the world.

The idea of the nose came out of a meeting in the early days of the charity, as the founders searched for a fun symbol that anyone could easily wear. "What about a red nose?" someone quipped.

The rest is history.

Since 1988, *Comic Relief* has become a fixture in the BBC calendar, alternating every other year with *Sport Relief*. By 2022, *Comic Relief* had raised over £1 billion pounds.

Red noses all *(from left)*: Claudia Winkleman, Alan Carr, Graham Norton, Davina McCall, Jonathan Ross, David Tennant and Fern Britton, 2009.

Noel's House Party

"Only those who've presented live television really know how difficult it is to make it look easy."

Noel Edmonds

First appearing in a 1992 "Gotcha" segment, Mr Blobby goes on to be a huge part of the show's success.

King of Saturday night entertainment in the 1990s is Noel Edmonds. Graduating from Radio 1 breakfast show DJ, via *Top of the Pops* (from 1972), *Multi-Coloured Swapshop* (1978) and *Telly Addicts* (1985), he creates the winning tea-time format of *Noel's House Party* in 1991.

A precariously live show, it is broadcast from an imagined mansion in the fictional village of Crinkley Bottom, and becomes hugely famous for its surprise celebrity visits, its gunge tank and "Gotcha!" moments revealed by hidden cameras. Finally, there's Mr Blobby, a large, inflatable, pink humanoid who only ever says "Blobby, blobby" – but in spite of that still manages to score the coveted Christmas Number 1 spot in 1993. The show has audiences of 15 million at its peak and lasts for eight years.

In this year

15 March
Comic Relief

15 April
Haute couture harks back to the 1920s with drama *The House of Eliott.*

3 September
Andrew Marshall's hit sitcom *2point4 Children* stars Belinda Lang, Gary Olsen and Julia Hills.

5 October:
Absolute Hell by Rodney Ackland is the first in a series of star-studded play adaptations to air on BBC Two under the umbrella title *Performance.*

23 November
Noel's House Party

It's his party – Pauline Quirke and Linda Robson from *Birds of a Feather* challenge Susan Tully and Michelle Gayle from *EastEnders.*

213

Later… with Jools Holland

"Without the piano my life would be a disaster – nobody would hold me in any regard. It's the thing that saved me."

Jools Holland

The perfect piano host
– Jools Holland in 2021.

Later… with Jools Holland comes quite literally after *The Late Show*. BBC Two Channel Controller Michael Jackson wants to exploit the empty studio space going unused after the latter arts programme, and he knows producer Mark Cooper is keen to create a new showcase for eclectic music. They also think that relaxed performer/presenter Jools Holland will be the perfect host, interspersing music with chat and his own bit of jazz piano.

And so begins *Later…* on 8 October 1992 and proves an instant hit. Audiences love it and so do the performing musicians. Virtually every style of popular music is featured and anyone who is anyone appears on the show – from Randy Newman, Smokey Robinson and Sting to Norah Jones, Coldplay and Adele – as well as many up-and-coming acts. There's an intimacy and a real attentiveness to the music. "I'm not sure if it comes across, but you can reach out and touch the band standing next to you," says singer Tracey Chapman.

Jools's Annual Hootenanny becomes a New Year institution, and *Later…* will go on to bring familiar and unfamiliar sounds to its loyal audience over the next decades. "Long may *Later…* reign," says Guy Garvey of Elbow.

Documenting Different Voices

"I adore this industry and believe ultimately in its power to do more good than harm, and its ability to create change. I for one, will not give up until disabled people get the equality in this industry we deserve."

Kim Tserkezie, presenter of *From the Edge*.

In 1973, the BBC created the Community Programme Unit, an innovative project to widen access to the airwaves for non-professional voices and campaigning groups. Out of it comes the extraordinary *Video Diaries* (1990), followed by *Video Nation* (1994), giving ordinary people – from football fans to prisoners – the opportunity to document their own lives in their own way, and transforming the documentary format as they do so.

In 1992, the Disability Programmes Unit is created, so that disabled people can speak for themselves. Its remit is diverse, from hard-hitting journalism to comedy commissions (agony aunt Daphne Doesgood), as well as the lifestyle strands *From the Edge* and *Feelings* (both 1995). There is also a BBC One commission, *The Invisible Wall* (1995), which uses secret cameras to show the shocking discrimination towards disabled people on the street.

The ground-breaking *Video Diaries* series allows people to document their own lives on air for the first time.

215

BBC Films

"A recognition of past and present achievements but also an affirmation of why BBC Films is so important for the film industry both here in the UK and further afield."

Chairman of BAFTA's Film Committee Nik Powell awards BBC Films the Outstanding British Contribution to Cinema, 2015.

A breakthrough, BAFTA-award-winning performance for Jamie Bell in *Billy Elliot*.

"I really, truly, madly, deeply, passionately, remarkably, umm... deliciously love you..." Nina (Juliet Stevenson) runs out of adverbs to Jamie (Alan Rickman).

Founded on 18 June 1990, BBC Films is the feature filmmaking arm of the BBC, created to tell great British stories for UK and global cinema audiences alike. Up to 2022, it will build an impressive catalogue of more than 300 films, from *Billy Elliot* (2000), *An Education* (2009) and *Pride* (2014) to *Notes on a Scandal* (2006), *In the Loop* (2009), and *The Power of the Dog* (2021).

Its founder is David M. Thompson who heads up BBC Films until 2007, to be followed by Christine Langan and then Rose Garnett (from 2017), all of whom reinvent the BBC brand in the complex and competitive world of international film.

Award winners

BBC Films quickly clocks up distinctive and popular successes, beginning with Anthony Minghella's memorable and emotive *Truly, Madly, Deeply* (1990), which combines an almost whimsical ghost story with a powerful portrayal of grief. It wins Minghella Best Screenplay – Original, and lead actors Alan Rickman and Juliet Stevenson are also nominated for BAFTAs for Best Actor and Best Actress respectively. The film gets its first television showing on 1 March 1992.

Other titles to capture the public's attention and awards recognition include Pawel Pawlikowski's *My Summer of Love* (2004); Andrea Arnold's *Fish Tank* (2009); *Philomena* (2013) directed by Stephen Frears; *Brooklyn* (2015), directed by John Crowley and written by Nick Hornby; Remi Weekes's *His House* (2020); Aleem Khan's *After Love* (2020); and Ken Loach's *I, Daniel Blake* (2016), which wins the Palme d'Or at the Cannes Film Festival.

Joanna Scanlan wins a Best Actress BAFTA in 2022 for her performance in the quietly revelatory and moving *After Love*.

Lashana Lynch in *ear for eye*, a blistering experimental film about British and American Black experience. She wins the Rising Star Award at the BAFTAs in 2022.

Oscars triumph also follows, for *Iris*, with Jim Broadbent named Best Supporting Actor, and *Judy*, for which Renée Zellweger is named Best Actress for her astonishing portrayal of Judy Garland during her last troubled series of UK concerts.

In 2020, the division is rebranded BBC Film, and BBC documentary strand *Storyville* comes under its remit. In 2021, the world premiere of *ear for eye*, (her name and film title always in lower case) filmmaker and playwright debbie tucker green's second feature, premieres at the BFI London Film Festival and exclusively the same evening on BBC Two and BBC iPlayer – a unique, multi-platform launch. And in 2022, Jane Campion becomes the third woman in history to win the Best Director Academy Award for *The Power of the Dog*, which also wins BAFTAs for Best Picture and for Best Director.

Kodi Smit-McPhee and Benedict Cumberbatch clash tragically in Jane Campion's "anti-western western" (*The New Yorker*), *The Power of the Dog*.

Absolutely Fabulous

> "There's normal life and then there's *Ab Fab*.
> And the division between the two is what's funny."
>
> Jennifer Saunders reflects on the comedy of *Absolutely Fabulous*.

**Joanna Lumley
(1946–)**

A model in the Swinging Sixties, Joanna Lumley's first acting job is on the 1969 James Bond film *On Her Majesty's Secret Service*. Later comes her breakthrough TV role in ITV's *The New Avengers* (1976). But it is her role two decades after that, as Patsy in *Absolutely Fabulous*, that completely reinvents her and makes her a global star. Drunk, disgusting, and merciless, Patsy remains a classic comedy character of the 1990s. Lumley goes on to present a range of documentaries including *Joanna Lumley's India* (2017), as well as lobbying passionately for the Gurkha Justice Campaign and Survival International.
She is made a Dame in 2022.

Originating from an earlier French and Saunders sketch about a clashing needy mother and her sensible daughter, *Absolutely Fabulous* blossoms into one of the huge comedy hits of the 1990s.

It arrives on our screens on 12 November 1992, starring Jennifer Saunders as self-absorbed fashionista Edina Monsoon, Julia Sawalha as her grounded, sensible daughter, and – in a stroke of inspired casting as she is not known for her comedy touch – Joanna Lumley as the outrageously debauched Patsy Stone, breaking every politically correct canon there is.

Sitcom royalty June Whitfield plays Edina's mother and Jane Horrocks completes the group as Edina's clueless PA Bubble. The show quickly becomes known as *Ab Fab* and Edina and Patsy become comic icons.

Significantly, there are no major roles for men, other than Edina's ex-husbands, Justin and Marshall, played by Christopher Malcolm and Christopher Ryan.

Ab Fab ends after five series, but its continued popularity gives rise to several specials and, in 2016, a film, *Absolutely Fabulous: The Movie* coruscating with star turns and celebrity cameos – an indication of the show's worldwide appeal. The show defines and sends up the excesses of the decade, and is yet another huge boost for female comedy.

> "Darling, if you want to talk bollocks and the meaning of life, you're better off just downing a bottle of whisky. At least that way you're unconscious by the time you start to take yourself seriously!"
>
> Joanna Lumley's Patsy

In this year

The Disability Programmes Unit (DPU) is created.

3 January
Laurence Marks and Marice Gran's comedy drama *Love Hurts* airs, starring Zoë Wanamaker and Adam Faith as lovers from very different walks of life.

22 March
Anthony Minghella wins a BAFTA for his screenplay for BBC Films' fantasy drama *Truly, Madly, Deeply.*

8 October
Later... with Jools Holland

12 November
Absolutely Fabulous

Women behaving badly! Patsy (Joanna Lumley) and Edina (Jennifer Saunders) hit the booze in 1996.

From The Multi-Coloured Swap Shop to Live and Kicking!

"The first show was in September 1996. I don't think I've ever been so terrified. I told myself I just had to get through the first five minutes. Then, the moment we went live, I managed to be looking at the wrong camera…"

Jamie Theakston recalls his first *Live and Kicking!* show.

Saturday mornings for children changed forever on 2 October 1976, when *The Multi-Coloured Swap Shop* (simplified later to *Swap Shop*) opened its fun-sized doors.

Hosted by TV's favourite presenter, Noel Edmonds, it was groundbreaking in so many ways – its Saturday morning takeover, its three-hour duration, and its innovative use of the phone-in format. Stuffed with competitions, music, cartoons, interviews with public figures, news coverage, as well as the hugely popular roving Swaporama giving kids the chance to swap things with other kids, it ran till 1982. Edmonds later recalled: "It was a programme that broke the mould and I am proud to have been a part of it."

To be succeeded by *Saturday Superstore* in 1982, similar in style and with some of the same presenters, plus featuring that famous telephone number: 01-811-8055 engraved indelibly on so many children's memories. Then five years later, Sarah Greene and Philip Schofield host *Going Live!*, offering a menu of pop music, gunge tanks, fun cookery demos, as well as the slightly more serious "Growing Pains" and "All About Me" sections, covering eating disorders, disability and even AIDS.

Live and Kicking!

On 2 October 1993, Andi Peters moves out of the continuity-themed Broom Cupboard to present the latest three-hour extravaganza for kids, *Live and Kicking!*, alongside Emma Forbes and rising star John Barrowman. Many favourite features continue, plus some new ones – "Famous for Five Minutes" allows a family the chance to shine; "It's My Life" looks at people with unusual lifestyles.

The show goes up a gear when presenters Zoe Ball and Jamie

Theakston take over in 1996, attracting 2.5 million viewers at its peak and winning a BAFTA award in 1999 for Best Entertainment Show. Ball will later go on to present *The Radio 2 Breakfast Show*, as well as *Strictly Come Dancing*'s aftershow *It Takes Two*.

"Actually I'm not sure who is at fault here, but something goes wrong and it's quite funny…" Zoe confesses all, while having a ball with Jamie Theakston in *Live and Kicking!*

The Fast Show

> "It was a case of 'Come on, say your line and get off!'"
>
> Charlie Higson on what defines *The Fast Show*.

"Suits you, Sir"; "Does my bum look big in this?"; "I'll get me coat"; "I'm a little bit wooh, little bit waaay". "...which was nice"; "I'm afraid I was very, very drunk"; and "Brilliant!..."

Fast by name, fast by nature, the catchphrases and sketches come thick and fast in this popular and effervescent show. According to creator Paul Whitehouse, the show's format comes from his experience of watching a five-minute reel of highlight clips ready to send to the BBC for preview, and he suddenly realizes that "character comes on, someone shouts 'ARSE!', bang, next sketch!" is actually funnier than the usual sketch build-up.

And so, Whitehouse and Charlie Higson build a show around that format. They are aided and abetted by a talented ensemble cast that includes John Thompson, Arabella Weir, Caroline Aherne, Mark Williams and Simon Day.

One of the *Fast Show*'s most interesting features is that as certain characters become established – in particular the drunken raconteur Rowley Birkin and the odd couple of posh landowner Ralph and monosyllabic estate worker Ted – they take on a life of their own, enabling subtlety and even poignancy to creep into the sketches.

The Fast Show launches on 24 September this year and runs for three series until 1997, returning for a one-off special titled *The Last Fast Show Ever* in 2000.

"I wouldn't know about that, Sir" – groundsman Ted (Paul Whitehouse) is baffled by the advances of repressed lord of the manor Ralph (Charlie Higson), 1997.

Clockwise from left: John Thomson, Paul Shearer, Paul Whitehouse, Simon Day, Mark Williams, Caroline Aherne, Charlie Higson and Arabella Weir, all having fast and furious fun in 1997.

Pride and Prejudice

> "One of the things I've always thought is a drag in so many period adaptations is that they are always buttoned up to the neck… I'm always looking for excuses to get them out of their clothes."
>
> Andrew Davies on his unstuffy approach to literary adaptation.

Andrew Davies (1936–)

Davies begins writing for TV in the late 1960s/70s, then in the late 1980s, starts to take on literary adaptations of modern texts including *Mother Love* (1989) and *The House of Cards* (1990). But it's his adaptation of George Eliot's *Middlemarch* (1994) followed by the huge success of *Pride and Prejudice* that mark him out as the classic text adaptor of the day, and lead to a prolific output – from *Vanity Fair* (1998) to an *EastEnders*-inspired *Bleak House* (2005) and *War & Peace* (2016). His work eschews textual faithfulness for modern reinvention, often revealing implicit sexual tension.

Just a matter of time before prejudice and pride are reconciled for Elizabeth Bennet (Jennifer Ehle) and Mr Darcy (Colin Firth).

Jane Austen's best loved novel, *Pride and Prejudice* has been adapted by the BBC for television at least five times, but this version, beginning on 24 September, signals a whole new approach to TV adaptation of the classics. The screenplay by Andrew Davies brings out the sexual charge only implicit in the novel, as well as creating many scenes not in the book – including the now-famous lake swim which involves Mr Darcy (Colin Firth) emerging dripping wet, diaphanous linen shirt clinging to his body.

Playing opposite Firth is Jennifer Ehle as a sprightly Lizzie Bennet, with Susannah Harker and Crispin Bonham-Carter as uncertain lovers Jane Bennet and Mr Bingley, Julia Sawalha as the flirtatious Lydia, and Alison Steadman and Benjamin Whitrow as the beleaguered Mr and Mrs Bennet. Producer Sue Birtwistle directs with verve and modernity, reflecting Davies' adaptation, while Dinah Collins' costumes are designed to look natural and wearable – unlike those in many costume dramas – allowing Lizzie to traipse the countryside and Lydia to dance giddily about the ballroom.

The six-part drama is a phenomenal success wherever it is shown. Its UK audience of ten million is captivated until the very end, while 100,000 video box sets fly off shop shelves. *Pride and Prejudice* reinvents the classic drama and establishes Davies as the go-to literary adaptor for the next two decades.

Filming the ball at Netherfield Hall.

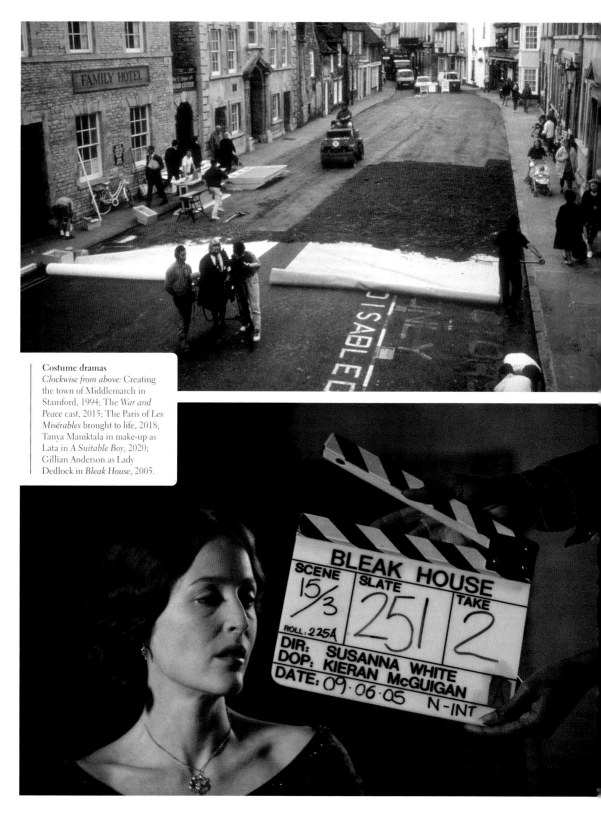

Costume dramas
Clockwise from above: Creating the town of Middlemarch in Stamford, 1994; The *War and Peace* cast, 2015; The Paris of *Les Misérables* brought to life, 2018; Tanya Maniktala in make-up as Lata in *A Suitable Boy*, 2020; Gillian Anderson as Lady Dedlock in *Bleak House*, 2005.

FAMILY HOTEL

DISABLED

BLEAK HOUSE
SCENE
15/3
SLATE
251
TAKE
2
ROLL. 225A
DIR: SUSANNA WHITE
DOP: KIERAN McGUIGAN
DATE: 09·06·05   N-INT

Silent Witness

> "If you like your crime dramas dimly lit, ably acted, only occasionally preposterous and chock-full of cut-open dead bodies, then *Silent Witness* is still right where it always was."
>
> *Guardian* review, 2016

Amanda Burton as Sam Ryan exploring the darker sides of pathology together with Leo (William Gaminara) and Harry (Tom Ward), 2003.

Created by former murder squad detective Nigel McCrery, *Silent Witness* opens its cadaverous proceedings on 21 February. It will run and run, eventually becoming the longest-running crime drama on BBC television – and anywhere in the world.

It focuses on a small team of forensic experts tasked with using their expertise to solve contemporary, often troubling and frequently gruesome, crimes. The show offers great leading roles for women – in particular Amanda Burton as pathologist Sam Ryan, to be succeeded by Emilia Fox as Nikki Alexander, and later – in one of the first mainstream roles for a disabled actor – Liz Carr as Clarissa Mullery.

From Do It Yourself to Changing Rooms

"I had only ever been designing swanky houses in Knightsbridge and Chelsea… Suddenly we were shoved into council houses and caravans and given £500. I loved the challenge of it."

Changing Rooms presenter Laurence Llewelyn-Bowen

Graham Wynne and Laurence Llewelyn-Bowen take on another *Changing Rooms* challenge, 1999.

Tapping into the post-war rebuilding ethos, DIY surges into prominence in the late 1950s and 1960s. *Do It Yourself* (1957) introduces us to Barry Bucknell who in 1962 presents *Bucknell's House*, a showcase for his "modernization" of an Ealing semi.

Then jump forward a few decades to the transformative mix of DIY and décor that is *Changing Rooms*. Debuting on 4 September this year, it turns viewers into quick-fix interior designers and runs for 17 series over eight years.

Presented by Carol Smillie, with interior designers Linda Barker, Anna Ryder-Richardson and Laurence Llewelyn-Bowen, plus carpenter "Handy" Andy, the show invites members of the public to redesign their neighbours' rooms. The audience's delight is both in the mismatch of what people imagine their neighbours' design tastes to be and the reality, as well as in the many disasters that happen along the way. The show runs till 2004, makes a star of the flamboyant Llewelyn-Bowen, and is recommissioned by Channel 4 in 2021.

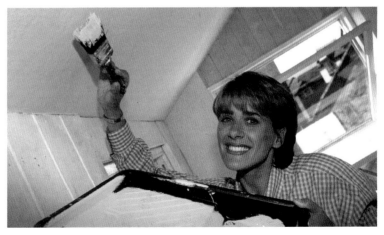

Presenter Carol Smillie helps out the DIY team in the long-running *Changing Rooms*.

Teletubbies

"Over the hills and far away, Teletubbies come to play!"

The opening words of each *Teletubbies* show.

Meet Laa-Laa: a bundle of energy and fun, she loves laughing and dancing, and her signature move is a light-footed twirl.

On 31 March, the Teletubbies arrive: Tinky Winky (big and purple), Dipsy (out there and green), Laa-Laa (shy and yellow) and Po (red and small). They are ready to sing and dance, eat tubby toast, play with Noo-Noo the vacuum cleaner, and take over the world!

Devised by Anne Wood and Andrew Davenport of Ragdoll Productions, *Teletubbies* is the natural successor to *Play School* (1964) and *Play Away* (1971), focusing on learning through play for pre-school children. It has a regular sequence of reassuring features, from its opening baby-faced sun to the TV inserts that appear in the characters' tummies. There is criticism early on that the show perpetuates baby talk and behaviour – particularly its endless repetition of "Eh-Oh!" as the characters bump up against each other – but much the same was said of Bill and Ben and their Flobadob language 40 years earlier.

However, children everywhere take the show to their hearts and it becomes a global phenomenon, selling in 120 countries in 45 languages by the end of 2002. Linked *Teletubbies* merchandizing sells, too, to a value of £120 million in the 1990s alone; the show even scores a Number One single in the UK charts. *Teletubbies* transforms the way that BBC Children's shows are developed and marketed, paving the way for *Tweenies* (2000), and *In the Night Garden* (2007).

The Teletubbies in the distinctive world of Teletubby Land.

bbc.co.uk and BBC News 24

"I know this is going to be an important public service medium…
and we've got to be there, and be pioneers."

Director-General John Birt assesses the importance of the worldwide web.

In this year

31 March
Teletubbies

28 April
The launch of BBC Online.

27 June
BBC broadcasts from the
Glastonbury Festival.

6 September
The funeral of Diana,
Princess of Wales is broadcast
live from Westminster Abbey.

9 November
Rolling news channel BBC
News 24 launches.

Towards the end of the 1990s, broadcasting begins a slow but seismic transformation… On 15 December this year, the BBC launches its very first website: www.bbc.co.uk. It is championed by Director-General John Birt, who can see how media will be transformed by the worldwide web, but he initially faces problems with its lack of inclusion in the BBC Charter – which still defines broadcasting solely in terms of radio and television.

Once the site is launched, other milestones quickly follow: the Queen's Christmas Speech goes online, and there are sites for sport and weather. By 2005, the BBC's online service is reaching almost 12 million users a month.

A month earlier, BBC News 24 begins, offering rolling news day and night, and meeting "an audience demand for news when they want it," as Tony Hall, Director of News, states. Slowly, a revolution is starting as the broadcast schedule is handed direct to the viewer and listener.

The constant evolution
of BBC News with its spinning
globe symbol, seen here in 2008.

Goodness Gracious Me

"Father Christmas? Indian!"

Mr "Everything comes from India" who is convinced that Superman, the royal family and Father Christmas are all Indian.

In this year

12 January
Goodness Gracious Me

14 September
The Royle Family

13 October
How to Cook with Delia Smith.

12 November
dinnerladies

The cast of *Goodness Gracious Me* (*left to right*): Nitin Sawhney, Sanjeev Bhaskar, Meera Syal, Kulvinder Ghir and Nina Wadia.

BBC Comedy shifts its perspective in the late 1990s, foregrounding new voices and new formats. At the top of the year is *Goodness Gracious Me*, which makes its TV debut on 12 January this year.

A sketch show like many others, it is unique in its cast being made up entirely of British Asians – igniting the careers of Sanjeev Bhaskar, Meera Syall, Nina Wadia and Kulvinder Ghir, along with producer Anil Gupta. As its title suggests, the show sends up Asian and British cultures, peppering its sketches with catchphrases – "Kiss my chuddies!" , "Rasmalai!" – along with a parade of stock characters.

Feel-good comedy enables the show to tackle playfully the serious topical issues of race, class and sex, and *Goodness Gracious Me* delights audiences for four series and then spawns the equally successful *The Kumars at No. 42* (2001).

From Wythenshawe to Royston Vasey

The Royle Family *(left to right)*: Ralf Little, Caroline Aherne, Sue Johnston, Ricky Tomlinson, Craig Cash and Liz Smith.

There's a bumper crop of divergent comedy sprouting across the BBC. *The Royle Family* – daughter Denise and fiancé Dave, son Anthony, parents Jim and Barbara, and Nana – are mainly slumped in front of their TV set in a front room in Wythenshawe, Manchester. Devised by Caroline Aherne and Craig Cash, the show takes the traditional domestic sitcom and strips away the laughter track to reinvent it as something altogether more real and character-based.

dinnerladies

Manchester, in the form of the fictitious canteen of HWD Components, is also the setting for Victoria Wood's much-loved *dinnerladies* (launched on 12 November and running for two series). It features Wood as Bren the deputy manager, trying to keep control of her unruly team of assorted workers and family members, including Julie Walters, Anne Reid, Thelma Barlow and Celia Imrie.

The League of Gentlemen

An altogether darker ensemble will hit TV screens next year, on 11 January, when *The League of Gentlemen* welcomes us to the inbred Northern gothic of Royston Vasey. Mark Gatiss, Steve Pemberton and Reece Shearsmith play all of the bizarre characters, as well as writing the series, along with fourth gentleman, Jeremy Dyson. Pemberton later comments: "That something so strange somehow found an audience on the scale it did was always the thing that amazed us all."

The gruesome welcome party of Edward (Reece Shearsmith) and Tubbs (Steve Pemberon).

WELCOME TO ROYSTON VASEY

YOU'LL NEVER LEAVE!

Walking with Dinosaurs

"Good fun in a carnivorous kind of way. Visually, it's a bobby dazzler."

The Guardian review of *Walking with Dinosaurs*.

In this year

The independently funded, international charity BBC Media Action (formerly the BBC World Service Trust) launches.

11 January
The League of Gentlemen

12 January
The first episode of medical drama *Holby City*.

12 April
Stop-motion animation series for children *Bob the Builder*.

14 April
The Naked Chef

6 September
Preschool live action series *Tweenies*.

4 October
Walking With Dinosaurs

"The undoubted television event of 1999," say the reviewers of *Walking With Dinosaurs*, the show which brings to life the fascinating world of prehistoric dinosaurs – for an audience already captivated by the 1993 Spielberg blockbuster *Jurassic Park*.

This six-part series extends the computer animation techniques developed for film, creating the first special-effects documentary of its type seen on the BBC, and indeed on British television. It is the brainchild of BBC science producer Tim Haines working with Mike Milne and his animation team, who between them manage to blend real-world filming with new and highly sophisticated animation, in order to conjure up before our eyes the world of 160 million years ago.

Pre-history brought thrillingly to life.

Outstanding animation

"Once our creatures were up and running, they looked magnificent and suddenly the era came alive – walking, talking, running, feeding and fighting, a whole menagerie of creatures, many of which have never been seen outside the pages of scientific journals," writes Haines. Underpinning the series is the question of why and how these ruling monsters became extinct.

With narration by actor Kenneth Branagh, music by Ben Bartlett, the show also has a bestselling book, soundtrack and immersive exhibition. It wins a plethora of awards, including two BAFTAs for innovation and two Emmys for outstanding animation and visual effects. The show creates a *Walking With…*franchise and is followed by the equally ambitious and enthralling *Walking With Beasts* (2001) and *Walking With Monsters* (2005).

Walking With Beasts explores the evolution of the human race.

The Naked Chef

"He hung out, did a dollop of this and a glop of that, and spoke like my family in Essex. It was a new kind of person and a new way of being around food."

BBC Two Controller Jane Root on first seeing Jamie Oliver on her channel.

A "pukka" dinner with Jamie!

TV cookery gets a lively jolt when Jamie Oliver hits our TV screens for the first time as *The Naked Chef* – "It's not me, it's the food!" he says, explaining the show's title, before speeding off on his moped to haggle for veg at London's Borough Market. Oliver's approach to cooking is a breath of fresh air after the careful cookery programmes of earlier decades, plus his shaggy-haired persona, easy manner and upbeat language – "Pukka! Pukka!" - endear him to a new generation of viewers. He makes three highly successful shows for the BBC, before moving to Channel 4 and launching major campaigns around healthy eating for all.

235

The 2000s
Life Online

The decade begins with the terror of 9/11 and a world torn apart.
We need to remind ourselves of the beauty of a blue planet, and
take comfort in human dexterity: dancing and baking. While
online timeshifts our lives and makes the unmissable unmissable.

Strictly judges Craig Revel
Horwood, Len Goodman, Alesha
Dixon and Bruno Tonioli get
their paddles out, 2010.

2000 Today

> "Twenty minutes into this, they announce that Putin has taken power in Russia."
>
> A viewer recalls the unplannable.

Shauna Lowry and Patrick Kielty get ready for the celebration.

In this year

1 January
2000 Today

18 January
Reality show *Castaway* follows a group surviving on a remote Scottish island.

26 March
Medical soap opera *Doctors*.

13 April
When Louis Theroux Met...

14 August
The Weakest Link

And so it finally happens…! We shift from one millennium to the next, our familiar dates change forever, and we wait for the Millennium bug to do its worst. Meanwhile, the BBC begins its 28-hour millennium broadcast, featuring stories, interviews and celebrations worldwide. The transmission is the longest live broadcast in the history of the BBC

"Hello from Greenwich..."
David Dimbleby opens the show as he walks along the meridian line. Other familiar presenters include Michael Parkinson, Gaby Roslin, Peter Sissons and Michael Buerk, plus many others popping up in various locations. It is a massive logistical feat, involving 60 broadcasters, headed by the BBC in the UK and WGBH in Boston, USA, while a wider editorial board involves ABC (Australia), CBC (Canada), CCTV (China), ETC (Egypt), RTL (Germany), SABC (South Africa), TF1 (France), TV Asahi (Japan), TV Globo (Brazil) and ABC (USA). The BBC provides the central production hub, receiving and distributing the 78 international satellite feeds required. *2000 Today* achieves a worldwide audience of 800 million.

The Weakest Link

On 14 August, rapid general knowledge quiz *The Weakest Link* begins its long run on our screens, fronted by "The Queen of Mean" Anne Robinson. Since 1993, she is known as a champion of consumer rights on *Watchdog*, but this show makes her a mistress of the put-down. Her dismissive "You are the weakest link, goodbye," as a failed quiz contestant leaves the show, is even used by Prime Minister Tony Blair in the House of Commons. The show will be revived in 2021 with a more benign host, Romesh Ranganathan.

Anna Robinson suffers no fools in *The Weakest Link*.

238

BBC News: 9/11

"I was on the ground floor of the South Tower when the first plane hit the North Tower with a sound [like] a huge container of concrete falling very near from a great height."

BBC Correspondent Stephen Evans from the World Trade Center.

"That perfect azure sky... like a Hollywood film set." BBC correspondent Stephen Evans recalls sitting in New York's World Trade Center on the morning of 11 September 2001.

In this year

9 July
The Office

11 September
The 9/11 terrorist attacks.

12 September
The Blue Planet

At 8:46 a.m. on 11 September, five al-Qaeda hijackers deliberately crash American Airlines Flight 11 into the northern facade of the World Trade Center's North Tower. At 9:03 a.m., another five hijackers crash United Airlines Flight 175 into the South Tower's southern facade. BBC Correspondent Stephen Evans is actually inside the World Trade Center as the events of 9/11 unfold, and will relive the terrible experience for years to come.

The horror of the terrorist attack will change the world forever, and it is also reported in a way that is inconceivable in previous decades – as BBC News covers the evolving story in detail on TV, radio and online, minute by minute, hour by hour.

Unfolding in real time

BBC News reacts quickly, producing 33 extra hours of news programming across all five radio networks, more than 40 hours of live rolling news for BBC World Service and an additional 14 hours of news on BBC One and BBC Two. This is supplemented by strong current affairs analysis, including *The World's Most Wanted*, *One Day of Terror for Correspondent* on BBC Two and a special *Analysis* edition for BBC Radio 4. Overall, during 11 September, 35 million people in the UK tune to BBC News on TV, radio and online.

Levels of audience interaction also reach an all-time high, with 15 million page impressions recorded on the BBCi News website on 11 September and a record 23 million the following day. Within just five days, 75,000 people email their thoughts and reactions as the world struggles to understand the most shocking images seen for a generation. BBC News 24 comes of age as national demand for the latest news reaches record levels. Significantly, the audience remains at a higher level into 2002 and beyond.

Top Gear

"Speed and power!"

Jeremy Clarkson on what drives him.

Strange for a show that began in April 1977 as a rather sedate motoring show, introduced by the *Radio Times* as the "first of a monthly series for road-users", that it can reinvent itself in 2002 as something altogether different.

Top Gear's first presenters are Angela Rippon and Tom Coyne. In the opening episode Rippon drives up the M1 in her Ford Capri – at a steady and safe 70 miles per hour - talking about the correct use of mirrors on the motorway. Later, the programme covers fuel efficiency and the introduction of seat belts and curbs on drink-drivers.

So a long way away from the show's radical re-hosting by the ebullient Jeremy Clarkson, Richard Hammond and Jason Dawe (later James May) on 20 October 2002, when it becomes a high-production, motoring-based entertainment show, featuring new segments such as "Star in a Reasonably-priced Car", "The Cool Wall", "Power Laps", and the hugely popular "Stig Test Drives"...

Jeremy Clarkson, a key player in the reboot of *Top Gear* in 2002.

Who is Stig?

The Stig is a racing driver whose identity is hidden behind a white tracksuit and helmet. He test-drives cars and sets lap times on the show's test track; Stig also preps celebrity guests for their drives. Named after the downtrodden character in the children's book, *Stig of the Dump*, the nickname comes from Clarkson's schooldays. Stig's identity has been constantly debated by fans, ranging from Formula One Champion Michael Schumacher to James May's mother!

Stig's madcap driving feats are just one of the iconic features that make *Top Gear* the BBC's most successful global brand, watched in over 250 countries. Clarkson leaves the show in 2015 following a bullying claim, and is replaced by a line of starry presenters, including Matt Le Blanc and Freddie Flintoff.

The anonymous and elusive Stig.

Louis Theroux Met...

"Documentaries can change the world."

Nick Fraser, founder of *Storyville*.

On the wider side of documentary making, Louis Theroux has been enlightening viewers with his unique and quirky style of film making since he came to prominence in the late 1990s. He's back on 13 April 2000 with a series called *Louis Theroux Met...*, beginning with his famous/infamous encounter with the DJ Jimmy Savile, followed by insightful meetings with Neil and Christine Hamilton, Anne Widdicombe and Max Clifford, among others.

Later this decade, the investigative documentary gets another shake-up with 2009's *Stacey Dooley Investigates...*, as Dooley tackles dark, sometimes dangerous issues, ranging from child prostitution and trafficking to female suicide bombers and digital drug dealers.

And from 1997 to 2022 and beyond, the flagship *Storyville* strand showcases the best in international documentary-making for the BBC, winning many awards and illuminating our lives.

Louis Theroux gets to know ex-Conservative MP Neil Hamilton ad his wife Christine in 2001.

Subversive Comedy: The Office, Little Britain, The Thick of It

"People see me, and they see the suit, and they go 'you're not fooling anyone', they know I'm rock and roll through and through."

The Office's David Brent

"Chilled-out entertainer" David Brent (Ricky Gervais) manages *The Office*, or thinks he does.

Subversive comedy begins in Slough, in the branch of the paper-supply company Wernham Hogg. First broadcast on 9 July 2001, written by Stephen Merchant and Ricky Gervais, *The Office* takes the mock-reality approach used in films such as *This Is Spinal Tap* (1984) and adapts it for the small screen. A pilot is shot in 2000, but the idea has been brewing with Gervais since the mid 1990s, based on a lot of bad work experience in his teens and twenties.

The show stars Gervais as ludicrously unself-aware office manager David Brent, joined by co-workers Tim Canterbury (Martin Freeman), Dawn Tinsley (Lucy Davis) and Gareth Keenan (Mackenzie Crook). After a slow start and some baffled critics, the show becomes one of the most influential comedies of the decade, leading to several international versions. And office life is never the same again....

Little Britain

Take a trip around *Little Britain* with David Walliams and Matt Lucas, and smash every taboo as you go. Shifting from radio to TV on 9 February, the show introduces us to a cast of vocal, sometimes grotesque, characters, including Bristolian juvenile delinquent Vicky Pollard ("No, but, yeah, but, no"); Daffyd Thomas, who claims to be the only gay in his village; Sebastian Love, government aide with a crush on the prime minister; and put-upon helper Lou and his wheelchair-bound friend Andy – but does he really need the wheelchair at all?

Little Britain is a massive success, with 9.5 million viewers following its move to BBC One in 2005; but owing to its host of highly controversial characters the series will be removed from BBC iPlayer in 2021 because "times have changed" (iPlayer wording).

Matt Lucas as Vicky Pollard, who has an answer for everything.

"You're so back-bench, you've actually f*****g fallen off. You're out by the f*****g bins where I put you."
Malcolm Tucker harangues some hapless MP.

Malcolm Tucker (Peter Capaldi) lets rip, while Jamie McDonald (Paul Higgins) listens in.

The Thick of It

Jump forward a couple of years to 19 May 2005, and comedy writer Armando Ianucci is having subversive (semi-improvised) fun, too.

The Thick of It satirizes the workings of modern British government, and the rise of the spin doctor. The best reason to tune in is to witness the insults of Malcolm Tucker, Peter Capaldi's foul-mouthed head of PR. The Thick of It spawns the Academy-Award-nominated film In the Loop (2009), written in direct response to the Iraq War.

QI

Sandi Toksvig, QI host from 2016.

Created by John Lloyd, the producer of radio's The News Quiz (1977), Blackadder (1983) and various other shows, QI begins life as an idea for an annotated encyclopedia game, mutates into a radio show, then, finally, on 11 September, is turned into a TV show – becoming a surprisingly long-running hit. The intellectual aspect of the show – embodied by the schoolmasterly manner of presenter Stephen Fry – is entertainingly tempered by the wisecracking comedian panellists, in particular regular Alan Davies, attempting to guess the answers, and usually failing spectacularly. The show will go on to win a clutch of awards, including RTS and British Comedy gongs.

In 2016, Stephen Fry passes the quizmaster seat to Sandi Toksvig. The BBC always takes Sandi at her word: "When people say, 'There aren't enough women on panel shows,' the answer is to make the host a woman."

The Iraq War Dossier

"A good piece of investigative journalism marred by flawed reporting."

Editor of *The Today Programme* Kevin Marsh's comment on Gilligan's interview.

News reporter Andrew Gilligan with revelations about a dossier.

The US-led invasion of Iraq to topple the country's dictator, Saddam Hussein, begins on 19 March this year, despite widespread public protests. On 29 May, John Humphrys is interviewing reporter Andrew Gilligan on the *Today* programme. The subject? The weapons of mass destruction that Saddam Hussein had, according to the government, stockpiled ready for deployment in 45 minutes. Gilligan asserts that a source has suggested to him that the government dossier outlining Hussein's weapons capabilities was "sexed up", in order to make the government's case for war more credible.

Protests are lodged by an incensed Alastair Campbell, the Labour Prime Minister Tony Blair's press advisor, who demands an apology. The BBC declares that it has nothing to apologize for. Then, in the following weeks, the name of the source is revealed to be the biological warfare expert David Kelly. He is obliged to explain himself publicly, and the strain of events leads to him taking his own life.

BBC Director-General Greg Dyke is mobbed by supportive BBC staff, following the announcement of his resignation in January 2004.

The gallery of News 24, during the publication of the Hutton Report, 28 January 2004.

The Hutton Report

In the wake of all this, the Government instructs Lord Hutton to conduct a public enquiry – witnesses to include the Prime Minister, the BBC's Chairman (Gavyn Davies) and Director-General (Greg Dyke), and senior MoD and security staff.

Hutton's 740-page verdict will be published in January 2004, and is "good news for the Government, and very bad news for the BBC", as one of the Corporation's reporters will state. Minimally critical of the government and highly critical of the BBC, it stuns the nation and leads to the resignations of both Gavyn Davies and Greg Dyke. The incident marks the lowest point in BBC relations with the Blair Labour Government.

Later, with a sense of more perspective, Gilligan's claim will look less contentious: more a clash between the quest for a scoop and the need for due process. But the damage is done, and the BBC must rebuild its top management as well as the morale of its News team.

Strictly Come Dancing

"A groundbreaking *Strictly* final in step with modern Britain."

The *Guardian* review of the 2021 final, which sees a same-sex couple and a deaf actor among the finalists.

Bruce Forsyth, original presenter of *Strictly Come Dancing*, tells the nation to "Keep dancing!"

Dance always had a key place in the BBC schedules, and ballroom dancing on television goes right back to the early days of the medium in 1937. Post-World War II, bandleader Victor Silvester's *Dancing Club* (1948) was a firm favourite, injecting a touch of much-needed glamour into austerity Britain – even though the glittering dresses described were only seen in black and white. It leads directly to the regional ballroom dance competition, *Come Dancing* (1950) which runs for over 30 years and will long be remembered for its formation dance teams, as well as the host of well-known presenters, including Terry Wogan, Angela Rippon, Judith Chalmers and David Jacobs.

The show is cancelled in 1998 to much public uproar, but after Baz Luhrmann's 1992 film *Strictly Ballroom* becomes a runaway hit, the BBC relaunches ballroom dancing with a difference – a celebrity dance contest called *Strictly Come Dancing* on 15 May. Ballroom is back with a vengeance…

"The good, the bad and the grumpy" – *Strictly*'s original judges, Bruno Tonioli, Arlene Phillips, Len Goodman and Craig Revel Horwood.

246

In this year

15 January
Starring John Hurt and Jenny Agutter, *The Alan Clark Diaries* dramatizes the memoirs of the controversial Conservative politician.

16 February
The Catherine Tate Show introduces viewers to a range of outlandish and eccentric characters, including swearing granny Nan, and stroppy teenager Lauren "Am I bovvered?" Cooper.

15 March
Strictly Come Dancing

20 April
Happy Birthday BBC Two celebrates the channel's 40th birthday.

23 June
Jed Mercurio's hard-hitting medical drama *Bodies*.

12 October
Celebrities explore their genealogies with often surprising results in *Who Do You Think You Are?*

Take your partners

Hosted by Light Entertainment legend Bruce Forsyth and Tess Daly, *Strictly Come Dancing* reignites Saturday nights on BBC One, becoming the surprise hit of the decade as the nation falls in love with dance all over again.

There are eight contestants in the first series, with newsreader Natasha Kaplinsky the eventual winner. As the show grows in popularity, so does the number of contestants, featuring an array of soap stars, actors, sports personalities, singers and youthful social media stars whose progress is avidly followed by the media and the press. The show's judges become star turns, too: at the start they are Len Goodman, Arlene Phillips, with "bad cop" Craig Revel Horwood and outrageous Bruno Tonioli, but later will include the ballerina Darcey Bussell and singer/past winner Alesha Dixon.

Strictly Come Dancing becomes a global hit: as *Dancing With the Stars*, it sells in well over 50 countries. In the UK the *Strictly* effect sees pair dancing – and dance in general – taken up with passion across the country. The show also plays a key role in reflecting social change, integrating disabled dancers and same-sex couples into the show's line-up, and making a big and popular statement about mainstream inclusivity.

2021 winners Rose Ayling Ellis (who is deaf) and Giovanni Pernice in their heart-stopping "silent dance" sequence.

Dragons' Den and The Apprentice

"Dragons' Den was the only window I had into business from 12 years old."

The newest Dragon in the Den, Steven Bartlett, who joins the show in 2022

The Dragons of 2021: Touker Suleyman, Sara Davies, Deborah Meaden, Tej Lalvani and Peter Jones.

In *Dragons' Den*, budding entrepreneurs get three minutes to pitch their ideas to five multimillionaires willing to invest their cash, time and expertise. The pitch is over when each of the Dragons declares "I'm out", or when the entrepreneur secures the investment that they are seeking. Presented by Evan Davis, the show's line-up of investors become household names: from Duncan Bannatyne, Peter Jones and Deborah Meaden to the latest and youngest Dragon, Steven Bartlett.

Dragon's Den features some inspired investments, including the Magic Whiteboard created by Neil and Laura Westwood and Levi Roots' Reggae Reggae sauce. There are also some jaw-droppingly awful moments when entrepreneurs freeze in front of the cameras…

Left to right: Baroness Karren Brady, Lord Sugar, Claude Littner.

The Apprentice

Entrepreneurship gets a second impetus this year from *The Apprentice*, Contestants battle through a series of business-related challenges, and each week one leaves the show with Lord Sugar's now-famous words "You're fired!" ringing in their ears. The winner gets a job at Sugar's company along with a chance to start their own business in partnership with Sugar, plus major investment funding. This popular, long-running show captures something of the post-millennium mood of mobility and change. It spawns a number of spin-offs, including comedic chat show *The Apprentice: You're Fired* (2006).

Springwatch

"A series that is able to capture and share the most dramatic and intimate
wildlife dramas as they unfold live."

Royal Television Society Craft and Design Innovation Award, 2016

Away from the cut and thrust of business, a close observation of nature
proves to be a new absorbing broadcast theme. *Springwatch* harks back
to a long tradition of BBC natural history programming begun in
post-war Bristol. Naturalist Desmond Hawkins launched a series of
programmes on the natural world, leading to the creation of the
Natural History Unit in 1957. Out of this will come the milestone
Attenborough series, but also *Springwatch* (launching on 30 May),
followed by *Autumnwatch* in 2006 and *Winterwatch* in 2012.

Largely hosted by Chris Packham and Michaela Strachan (taking
over from Kate Humble in 2012), along with various guest presenters,
all three series drive technological innovation in live broadcasting to
allow viewers to enjoy precious glimpses of previously secret animal
lives, and celebrating the amazing and varied wildlife and habitats of
Britain. Different media platforms also come together on these series:
TV, Red Button, iPlayer, online and social media, while collaborations
with CBBC, CBeebies, check-ins on BBC Radio 2, 4 and 6 Music, as
well as live-event link-ups, enable *Springwatch* to reach as wide an
audience as possible.

The *Springwatch* presenter team:
Gillian Burke, Chris Packham,
Michaela Strachan, Martin
Hughes-Games, 2017.

Life on Mars

"*[Life on Mars* is] *The Bill* x *Doctor Who* = *The Sweeney*, if you're looking for a mathematical equation to sum it up."

The Guardian

Is he mad or in a coma… or has detective Sam Tyler really travelled back in time? To the soundtrack of David Bowie's eponymous song, *Life on Mars* responds to post 9/11 uncertainties by taking us back to the 1970s so that a 2006 crime can be unpacked and resolved.

Created by Matthew Graham, Tony Jordan and Ashley Pharoah, the drama is an ingenious mix of nostalgia and innovation with stand-out roles for John Sim (as Tyler) and Philip Glennister (as sexist, old-school copper DCI Gene Hunt). It's an art director's dream, too – where everyone wears brown flares and shirts with enormous collars, where the wallpaper is an eyeball-jazzing nightmare, and where the ubiquitous car is a Ford Cortina Ghia. The show spawns a sequel, *Ashes to Ashes* (2008) plus an American adaptation.

The One Show

"As soon as you've got a Hollywood guest talking about things that people at home care about, that's when the show really works."

One Show presenter Matt Baker

How to fill that difficult TV slot between the end of daytime and the beginning of evening viewing? That's the dilemma for BBC TV schedulers. Initially, there is the "Toddlers' Truce", when television screens went blank for an hour while mothers (it was always presumed to be mothers in those days) put the children to bed.

When that system goes, in February 1957, the BBC conceives *Tonight*, a current-affairs show presented by Cliff Michelmore; then, on 9 September 1969 *Nationwide* launches. As its name suggests, this magazine-style programme, presented by Frank Bough and

Cliff Michelmore presents *Tonight* which runs from 1957 to 1965.

Michael Barratt, attempts to link all the BBC regions. Unfortunately, at times, Newcastle might be heard but not seen, Manchester might be seen but not heard, while Bristol might disappear altogether...

One for all

And then on 14 August this year, *The One Show* begins, featuring presenters Adrian Chiles and Nadia Sawalha as the first residents on the programme's brightly coloured couch. They are followed by a succession of TV personalities, including Mylene Klass, Christine Bleakley, Chris Evans, Matt Baker, Alex Jones, Ronan Keating, Lauren Laverne and Jermaine Jenas.

It's a live show that mixes topical features with big-name studio guests, and, like *Nationwide*, it strives to pull in the whole of the UK with stories large and small, from floods, riots and referenda to royal weddings and births. *The One Show* regularly draws in five million viewers for each programme every week.

In this year

9 January
Life on Mars

9 March
School drama *Waterloo Road*.

14 August
The One Show

14 September
Sketch show *That Mitchell and Webb Look*.

4 December
The Choir's Gareth Malone gets the UK singing.

Make a daily date with Alex Jones and Matt Baker on *The One Show*, 2014.

The Graham Norton Show

Graham Norton
(1963–)

Graham Norton trained as an actor, and starred in the Channel 4 comedy *Father Ted* (1995) before branching into mainstream presenting. His long-running BBC chat show is a key anchor of the BBC One schedule, plus he fronts a wide range of entertainment programmes, such as talent show *Any Dream Will Do* (2007). In 2009, Norton hosts the Eurovision Song Contest following in the steps of fellow Irishman Terry Wogan and, from 2 October 2010, gains his own Saturday morning Radio 2 show.

On 22 February, Graham Norton hosts his first BBC chat show – to run for an extraordinary 30 seasons and still running, picking up five BAFTA awards along the way.

Starting every show with his trademark jokey monologue, Norton then invites the pick of the day's glitterati – UK and Hollywood – onto his sofa. The conversation is witty, fluent, full of banter; his gift is to make it all seem so easy, while allowing the unusual and the revelatory to slip out and often to happen.

Award-winning Graham Norton, who kickstarts the chat show and makes it an appointment to view.

The roll call of celebrity guests is seemingly never-ending – from Tom Hanks and Judi Dench, to Penelope Cruz and Daniel Craig – and yes, they may have a film to promote or a book to sell, but the conversation is still more interesting than mere self-promotion. As one reviewer says of the show's American relay: "Graham Norton shoulders the burden of keeping the raucous cocktail party on his couch lively, and he does so masterfully."

Memorable moments

"Raucous" moments include Hugh Jackman and Bill Crystal reading *Bake Off* innuendos in front of Mary Berry; John Boyega and David Beckham fighting with *Star Wars* lightsabers; Emma Stone (almost) meeting the Spice Girls; and Hugh Grant reprising his famous romantic speech from *Four Weddings and a Funeral* to a dripping-wet Norton. Audiences get to be involved, too, via the infamous red chair, where a member of the public tells a story and has to get a laugh, otherwise they are tipped into backstage oblivion.

BBC iPlayer

"Making the unmissable, unmissable"

Marketing slogan for BBC iPlayer.

By 2012, BBC iPlayer is a leading media brand, and combined with BBC Online creates a BBC powerhouse – as exemplified in its showcase of the London 2012 Olympics, offereing, 24/7 coverage of every sport in the competition for the first time ever.

On 25 December, the world of broadcasting changes forever. Certainly not realized at the time, by the next decade television and radio schedules look irrelevant as viewers and listeners themselves choose what and when they watch or listen to… It's the age of iPlayer.

The idea of "video on demand" has been around a while – certainly talked about in the 1990s. In 2001 Radio Player leads the way, time-shifting radio content so programmes can be listened to via computer.

Television, however, is a different matter. There's the complicated issue of rights, and of appropriacy of content with no watershed to control it. After two years of internal presentation and external debate, My BBCPlayer finally gets the go-ahead, with an initial limited shop window of 30 days to view. By 2012, it is a leading media brand.

Gavin and Stacey

"What's occurring?"

The famous catchphrase of Nessa (Ruth Jones).

When Gavin meets Stacey (Mathew Horne and Joanna Page), along with *(from left)* Bryn (Rob Brydon), Gwen (Melanie Walters), Nessa (Ruth Jones), Pam (Alison Steadman), Mick (Larry Lamb) and Smithy (James Corden), 2007.

James Corden comes up with the idea for *Gavin and Stacey* at a wedding reception, when he remembers that his best mate Gavin met his future wife when she phoned up his work… And so the seed is sown for the slow-burn comedy that becomes the nation's favourite. It is co-written with actor friend Ruth Jones, and both Corden and Jones also star as oddly matched couple Smithy and Nessa, best friends of Gavin and Stacey (Matthew Horne and Joanna Page), and part of the extended Essex/Barry Island family that includes such top character-acting talent as Rob Brydon, Alison Steadman and Larry Lamb. At first Corden and Jones only intend Gavin and Stacey to be a one-off, but the BBC commissions a series, starting on 13 May.

The show becomes a huge success, moving from BBC Three to BBC Two to BBC One. A 2019 Christmas Special attracts 11.6 million viewers and is the most popular scripted show of the year.

Outnumbered

"It was a real assault course for the actors. It was like being a real parent. You don't necessarily know what trouble the kids are going to throw up."

Outnumbered writer Guy Jenkin

The adage "never work with children or animals" is turned on its head in the deft hands of comedy writers Andy Hamilton and Guy Jenkin. *Outnumbered* puts three young children at the centre of this family-based sitcom and allows them to improvise into the bargain.

Hamilton and Jenkin worked together on Channel 4's highly successful *Drop the Dead Donkey* (1990), but this show is genuinely experimental: "So there is a script, but we never show it to the children and they never learn their lines…" says Hamilton.

Battling gamely with the improvisations of the children (Tyger Drew-Honey, Daniel Roche and Ramona Marquez) are Mum and Dad (Claire Skinner and Hugh Dennis) with drop-in visitors Grandad (David Ryall) and Auntie Angela (Samantha Bond). *Outnumbered* runs for six successful series till 2014, plus a Christmas special in 2016 – by which point the children have truly outnumbered the adults by growing up!

Beleaguered parents Claire Skinner and Hugh Dennis are outnumbered by Ramona Marquez, Tyger Drew-Honey and Daniel Roche.

A Matter of Loaf and Death

"Over the years the BBC has been incredibly supportive of Wallace and Gromit, this film feels like their homecoming."

Nick Park, Aardman Animations

In this year

The World Service launches new multimedia Arabic and Persian Services.

10 March
BBC Film *White Girl* follows a white family in Leeds who move into an area of the city mainly inhabited by people from the South Asian community.

28 May
The life of "moral guardian" Mary Whitehouse, who frequently clashed with the BBC in the 1960s and 70s is dramatized in *Filth: The Mary Whitehouse Story*.

30 July
The rise and fall of the Iraq dictator Saddam Hussein is dramatized in *House of Saddam*.

24 August
Alexander Armstrong and Richard Osman test contestants' obscure knowledge in *Pointless*.

14 September
Thomas Hardy's novel *Tess of the D'Urbevilles* stars Gemma Arterton.

24 October
Charles Dickens' *Little Dorrit*, stars Claire Foy.

Up to Gromit as ever to save the day, with Wallace and Piella completely out of control.

Wallace and Gromit open a new bakery – called Top Bun (of course!) and business is booming, not least as a deadly "cereal killer" has murdered all the other bakers in town. Gromit is worried, but Wallace doesn't care as he has fallen head over heels in love with Piella Bakewell, former star of the Bake-O-Life commercial… It all adds up to a classic "who-doughnut" mystery from the master hand of Nick Park's Aardman Animations, which is BBC One's Christmas Day treat of 2008. A *Matter of Loaf and Death* goes on to win the BAFTA for best animated film and is also nominated for an Academy Award.

Shaun and friends

A year earlier, on 5 March 2007, Aardman's *Shaun the Sheep* launched himself on CBBC's immediately appreciative audience. Shaun initially featured in the film *A Close Shave* (1995), and he will go on to conquer the small screen in more than 180 countries, also inspiring Christmas TV specials and two feature films. Finally, Shaun is voted the nation's best-loved BBC Children's character in a 2014 *Radio Times* poll.

Horrible Histories

"The Georgians with their big hair and make-up are very popular, but, from a pure comedy point of view, you really can't beat the Romans."

Richard Bradley, Executive Producer, *Horrible Histories*

History was never so horrible nor so much fun till now...

In this year

25 January
Cerrie Burnell is the first disabled presenter on children's channel CBBC.

16 April
Horrible Histories

9 November
Miranda

Based on Terry Deary's bestselling books, *Horrible Histories* debuts on CBBC on 16 April, satirizing historical events, and adding true facts to delight children and adults alike. However, what really makes the series stand out are the top comedy writers, such as Steve Punt, Jon Holmes (*Dead Ringers*) and Giles Pilbrow (*Have I Got News For You*) and actors such as Meera Syal (*Goodness Gracious Me*), Jim Howick (*Peep Show*), Simon Farnaby (*The Mighty Boosh*), Mathew Baynton (*Gavin and Stacey*) and Ben Ward (*Dead Ringers*). Dominic Brigstocke, director of *Green Wing* (2004) and *The Armstrong & Miller Show* (2007), also plays a key role. In 2021, *Horrible Histories* wins Best Sketch Show as voted by the public.

Miranda

"As a woman, it seems you can't just be a comedian; you're always classed as something else, too, whether that's 'beautiful,' 'pint-sized,' 'larger-than-life' or in my case, 'Hattie Jacques-esque,' 'the giraffe,' 'big'..."

Miranda Hart

Award-winning Miranda makes high-flying comedy out of mishaps.

Miranda Hart in *Miranda* hits our screens on 9 November. Old-fashioned though it is in many ways – the girl who can't get the man – the self-deprecating persona and comedic timing of both Mirandas make a winning formula. Her comic inspiration is Eric Morecambe (she presents the show *My Hero, Miranda Hart on Eric Morecambe* in 2013), in particular his trademark asides to the camera. Supported by a cast that includes Patricia Hodge, Tom Ellis, Sarah Hadland and Sally Phillips, the show wins every comedy award going.

The 2010s
Brand GB

What is this country? Polarized Britain surprises itself with a glorious broadcast Olympics, then severs itself from Europe. Meanwhile, we seem to rediscover the sky at night, and David Attenborough reflects back to us a world we must save – and fast.

Welcome to the opening night
of London 2012 Olympics in
the Olympic Park, East London,
27 July 2012.

Sherlock

"I'm not a psychopath... I'm a high-functioning sociopath. Do your research."

Sherlock's riposte to fellow detectives trailing in his wake.

Sherlock and Watson (Benedict Cumberbatch and Martin Freeman) in pursuit of fiendish criminality.

Sherlock is a deft, contemporary retelling of the adventures of Sir Arthur Conan Doyle's detective hero Sherlock Holmes. Devised and written by Steven Moffat and Mark Gatiss (who also appears as Sherlock's brother Mycroft), it hits our TV screens on 25 July and runs for seven years. It is a perfect example of the BBC taking a British classic and reinventing it for new and global generations, while still keeping the spirit of the original.

In this contemporary version Dr John Watson (Martin Freeman) is a war veteran home from Afghanistan, while Benedict Cumberbatch reinvents Sherlock as a mentally forensic but emotionally perplexed problem solver, or as *The Guardian* calls him, "cold, techie, slightly Aspergerish".

As soon as Watson moves into Holmes' Baker Street flat under the watchful eye of Mrs Hudson (Una Stubbs) they are embroiled in a series of mysteries, and Sherlock's nemesis, Moriarty (the mercurial Andrew Scott) appears to have a hand in the crimes...

Global hit

The show cleverly integrates new technologies on screen, and spices its new characterizations with a winning irony. It becomes an international phenomenon, with a huge following in China, where they call Sherlock "Curly Fu", and in Japan, where fans pore over *Sherlock*-inspired manga.

Luther

"Luther… represents Elba at his absolute best. It's a masterpiece of a performance, heavy and soulful and littered with all kinds of unexpected choices."

The Guardian

Devised by writer Neil Cross, *Luther* has – like *Sherlock* – a troubled, complex man at its core. Indeed, Cross compares John Luther to another iconic BBC character: "Its inspiration is not that different from Doctor Who, as in both cases you've got a trickster figure who fights the monster of the week and is eventually successful." Launching on 4 May and developing an avid following, the show follows the trials and misdemeanours of a near-genius detective whose brilliant mind can't always save him from the violence of his passions.

The aesthetic of the show is very British, filmed in the dark cityscapes of Central and East London, in such distinctive locations as the all-concrete Barbican Estate, Liverpool Street station, Roman Road Market and Docklands. As troubled DCI John Luther, Idris Elba delivers a dominant, driving performance as he tracks serial killers in the capital while desperately seeking to avoid being consumed by the darkness of their deeds. *Luther* sells in over 200 territories, and wins a clutch of national and international awards, including a Golden Globe for Best Actor.

"John is so close to my heart, he's a part of me," says Elba of the role he will play for almost a decade.

Home Comforts: The Great British Bake Off and Sewing Bee

"You've got too much of a soggy bottom."

Mary Berry's famous unintentional innuendo.

In this year

24 January
John Sullivan's *Rock & Chips*, a prequel to his *Only Fool and Horses*.

4 May
Luther

25 July
Sherlock

17 August
The Great British Bake Off

Star baker Mary Berry, helping the nation fall in love with baking all over again.

"Isn't competitive baking a contradiction in terms? It sounds as antithetical as competitive reading or competitive bath-soaking," ponders Guardian critic Lucy Mangan when *The Great British Bake Off* launches on 17 August.

And yet, take one tent and fill it with 12 amateur bakers, add an esteemed cookery writer and a soon-to-be famous bread-maker, along with a duo of comedians and a dollop of innuendo, and you have the surprise hit of the decade. The show makes baking a national passion, shifts from BBC Two to top evening billing on BBC One, and ends up in a pitched battle with Channel 4 for takeover.

Nadiya Hussain and some summer feast, 2019.

"On your marks, get set, 'Bake!'"

The inspiration behind *Bake Off*, as *The Great British Bake Off* soon becomes known) is American, from a friend of executive producer and creator Anna Beattie (of Love Productions) who has seen "bake-offs" in the US. Ironic then that the show will be exported to the USA.

The show makes stars of judges Mary Berry and Paul Hollywood, consolidates the ubiquity of presenters Sue Perkins and Mel Giedroyc, and launches the stellar career of 2015 winner Nadiya Hussain. It also leads to spin-off *Junior Bake Offs* on CBBC from 2011. In 2016, after much media speculation, the show moves to Channel 4 for a price of £75 million for a three-year contract.

Sustainable values

Off the back of *Bake Off* comes a raft of similar reality shows, delving deep into old-fashioned, home-maker skills. They are part of a gradual shift away from the all-pervasive materialism of the end of the last century towards values that are simpler and more sustainable.

Hugely popular is *The Great British Sewing Bee*, airing on BBC Two from 2 April 2013, presented by Claudia Winkleman, with a pair of judges, Patrick Grant and May Martin (then Esme Young). Winkleman is succeeded by Joe Lycett then Sara Pascoe, and celebrity versions are created for Children in Need.

> "I think you have to just put this one to the back of your mind… and the back of the wardrobe."
> *The Great British Sewing Bee*'s Patrick Grant to a hapless contestant.

Sales of sewing machines and sewing patterns go through the roof, as everyone catches the sewing bug.

The Sky at Night and Stargazing Live

"The darkness is full of wonders if you raise your gaze above the horizon."

Professor Brian Cox

**Brian Cox
(1968–)**

Frequently characterized as "the pop idol who became the science idol", because of his success as a keyboard player in pop bands of the late 80s and 90s, Cox translates his passion for music into a passion for communicating science. He is Professor of Particle Physics at the University of Manchester and importantly The Royal Society Professor for Public Engagement in Science. On the BBC, he inspires interest in the topic through a raft of programmes such as *Wonders of the Solar System* (2010), *Wonders of the Universe* (2011) and *Wonders of Life* (2012), *Stargazing Live* and *The Planets* (2019).

BBC Television's love-affair with astronomy goes back to 24 April 1957 and the first edition of *The Sky at Night*, consolidating earlier one-off programming on the subject. Its host is Patrick Moore, who presents the programme until his death in December 2012 – making him the world's longest-serving presenter of the same TV programme. Moore's expertise, enthusiasm and eccentricity become inextricably linked to astronomy.

Enter Brian Cox

Brian Cox, who will become the Patrick Moore of his generation through programmes such as *Stargazing Live*, credits Moore with inspiring his love of astronomy, when he presented him with a school prize: *The Observer's Book of Astronomy*! Cox's *Stargazing Live* show captivates audiences with astronomy in action. Across its regular annual slots, it chases the Northern Lights from an aeroplane above the clouds, observes the globe in a total solar eclipse, and celebrates British astronaut Tim Peake's launch into space.

Stargazing Live is a complex partnership with international space agencies, academics and some of the most famous astronomy facilities in the world, such as Jodrell Bank Observatory. In 2014, the show wins the Broadcast Digital Award for Best Content Partnership.

Stargazing Live's spin-off show, *Eclipse Special*, gains 4.2 million viewers on BBC One Daytime, and a clip from the broadcast on the BBC One Facebook page reaches 45.4 million global users and receives 12 million views.

"[Moore's] impact on the world of professional astronomy as well as amateur is hard to overstate."
Dr Marek Kakula, Public Astronomer at the Royal Observatory, Greenwich.

The Scandal of Jimmy Savile

"On behalf of the BBC and its staff past and present, I want to apologize to the survivors for all they have suffered. I also want to commit to them directly that we will ensure the BBC does everything it possibly can to prevent any such events in the future."

Rona Fairhead, Chair of the BBC Trust, on the publication of Dame Janet's report, 25 February 2016.

In this year

3 January
Stargazing Live

13 January
Documentary series *Human Planet* explores the many ways human beings interact with their environment around the world.

31 January
Stop-motion children's show *Rastamouse* features crime-fighting mouse reggae band Da Easy Crew.

21 February
Mrs Brown's Boys

29 April
The Wedding of Prince William and Kate Middleton.

29 October
The death of DJ and presenter Jimmy Savile.

When the DJ and presenter Sir Jimmy Savile dies on 29 October, his golden coffin is laid in state for 4,000 fans in his home town of Leeds. Described as "a legend" for his charity work, a year later he is a discredited sex offender and the revelations of his crimes rock both the BBC and the NHS. Evidence comes to light of his serial abuse at both the BBC and multiple medical institutions, including Leeds General Infirmary and Broadmoor Hospital, where he was ostensibly involved in charity work. Posthumously, he is found to have sexually abused 57 women or girls and 15 boys at the BBC, dating back to 1959.

The Savile crisis leads to the resignation of Director-General George Entwistle in November 2012 after only 54 days in the job. He has had to grapple with controversy over two *Newsnight* programmes, one on Savile that was shelved at the time of his death, and one that wrongly accuses Tory Lord McAlpine of child abuse.

The fall out

In the same year, the BBC asks former High Court judge Dame Janet Smith to conduct an independent review examining both the historical atmosphere at the BBC that enabled such behaviour, as well as making recommendations for conduct in future. She accepts that the BBC had no specific knowledge of Savile's horrendous crimes, but defines the BBC of the period when Savile's star was in the ascendant as a "culture of not complaining". It was in thrall to the power of television celebrities who were "handled with kid gloves" and were "virtually untouchable".

Dame Janet Smith presents the results of her report, February 2016.

London 2012

"The greatest day of sport I have ever witnessed!"

Lord Coe on "Super Saturday".

London Bridge lit up for the Opening Night.

"It was the best of times, it was the worst of times." The BBC's Chairman Lord Patten quotes the famous opening of Charles Dickens' *A Tale of Two Cities* in his introduction to the BBC's Annual Report 2012/13.

The success and quality of the BBC's Olympics coverage is rapidly followed by some of the darkest days in the Corporation's history, as revelations about the appalling crimes of DJ and presenter Jimmy Savile come to light.

But in the summer of 2012 the BBC's Olympics coverage, like the Games themselves, surpasses all expectations. As Danny Boyle's mesmerizing opening ceremony unfolds and as the Queen meets James Bond (Daniel Craig), it becomes clear that this is going to be a moment that brings the nation together and will long hold its place in our collective memory.

London 2012 Opening Ceremony, on 28 July.

Ben Brown reporting from the news studio at the heart of the Olympic Park.

Medals galore

There are so many great sporting moments – Mo Farah's 10,000-metres triumph, a shower of gold medals, including those won by track cyclists Bradley Wiggins and Chris Hoy; boxers Nicola Adams and Katie Taylor; Charlotte Dujardin for dressage; and sailor Ben Ainslie on the water. On "Super Saturday", track and field athletes Jessica Ennis-Hill, Greg Rutherford and Mo Farah all win gold in an unforgettable 44 minutes…

Billy (Perry Fenwick) from *EastEnders* takes the torch home to the Olympic Park.

Blanket coverage

The Games are illuminated by presenters Clare Balding, Gabby Logan, and Gary Lineker, and supported by experts including Denise Lewis and Michael Johnson. London 2012 is also the first truly digital Olympics with 24 streams of event coverage, meaning audiences don't "miss a moment". Every sport from every venue is covered, on TV and radio and online.

A grand total of 52 million people in the UK watch the London 2012 Olympics on the BBC.

Call the Midwife

"Our core values of passion and compassion never change, but the fact that time moves forward keeps the drama feeling fresh."

Call the Midwife writer Heidi Thomas

Call the Midwife soon grows a comitted regular audience of over 10 million. It is also an international hit.

In this year

15 January
Call the Midwife

24 March
Seeking unsigned talent, singing competition *The Voice* airs, judged by a panel of top recording artists.

26 June
Line of Duty

27 July
The London Olympics

Based on the memoirs of nurse Jennifer Worth and written by Heidi Thomas, *Call the Midwife* launches on 15 January. It follows 22-year-old Jenny (Jessica Raine), who in 1957 leaves her comfortable home to become a midwife in London's poverty-stricken East End, and is surprised to find herself in a convent working alongside fellow nurses and medically-trained nuns.

The drama has "that Sunday evening slot and good-old-days period setting" (*The Guardian*), but underneath its cosy exterior tackles some hard issues – from backstreet abortions and the thalidomide scandal to depression, homosexuality and racism. As reviewer Caitlin Moran writes: "You dilate those cervixes without the help of pethidine on prime-time BBC One... I raise my glass... to you!"

With a starry cast including Vanessa Redgrave, Jenny Agutter, Judy Parfitt, Stephen McGann, Helen George and Miranda Hart, *Call the Midwife* runs for the whole decade – and beyond.

Line of Duty and Bodyguard

"For something like Line of Duty to work, it has to be both plausible and unexpected."

Jed Mercurio

Line of Duty's core leads: Vicky McClure (DI Kate Fleming), Adrian Dunbar (Supt Ted Hastings), and Martin Compston (DS Steve Arnott).

It's the decade of screenwriter Jed Mercurio. He has previously had a hit with the RTS-winning medical drama *Bodies* (2005), but is now searching for a new way to write about the police: "We were looking at police misconduct, taking real-world examples to allow us to construct something that felt credible and relevant." The result is *Line of Duty*, which debuts on BBC Two on 26 June, and will become "The most-watched UK drama of the 21st century", according to *Radio Times*.

The story focuses on young DS Steve Arnott transferred to the anti-corruption squad to work with DCI Tony Gates (Lennie James), recently awarded a medal but with a suspiciously high arrest rate... The narrative device of a corrupt officer under investigation by AC-12 (anti-corruption investigative unit) is the show's irresistible dynamo. In subsequent series, the equivalent part to Gates will be played by Keeley Hawes (DI Lindsay Denton); Daniel Mays (Sgt Danny Waldron); Thandie Newton (DCI Roz Huntley); and Stephen Graham (DS John Corbett).

Bodyguard

Mercurio will return in 2018 with the hugely successful one-off drama *Bodyguard*, starring Richard Madden as a troubled war veteran David Budd, now a sergeant in the Metropolitan Police's Protection Command, assigned to guard Home Secretary Julia Montague (Keeley Hawes). The drama mixes Mercurio's trademark political suspense with sexual tension. As one critic says of Madden: "There's 50 shades of 'ma'am' in the way he addresses her."

Richard Madden stars as the brooding *Bodyguard*.

The Dumping Ground

"Tracy Beaker was my biggest inspiration growing up… thank you Tracy, for all the fabulous memories!!"

A CBBC Tracy Beaker fan

The cast of *The Dumping Ground*.

Based on Jacqueline Wilson's much-loved books, *The Dumping Ground* (15 March) is the last show made by CBBC about Tracy Beaker in the Elm Tree House care home. The show goes back to 2002, when rebellious foster child Tracy (Dani Harmer) first burst onto CBBC screens and became a runaway hit.

The various *Tracy Beaker* shows reflect the boldly child-centred world of the CBBC channel. They touch on difficult themes – bullying, self-image, identity, loss – as Tracy navigates the challenges of growing up in care, and the heartbreak of fantasizing that her mum, supposedly a Hollywood superstar, will one day whisk her away. But Tracy always wins through – usually with her catchphrase: "Bog Off!"

Drama meets fact when *Newsround* follows the lives of real children in the 2015 documentary, *The Real Tracy Beaker*.

My Mum Tracy Beaker

"Bog Off!" says young Tracy Beaker (Dani Harmer) in 2002.

Tracy Beaker returns in 2020 as an adult in the drama *My Mum Tracy Beaker*. Tracey is now a single mum to her 12-year-old daughter Jess (Emma Davies); life is still going "classically Tracy Beaker pear-shaped"; and it's Jess who has to save the day. The show breaks records as the most-streamed CBBC programme of all time, and leads to a 2021 follow-up, *The Beaker Girls*.

Peaky Blinders

"I don't pay for suits. My suits are on the house or the house burns down."

Tommy Shelby (Cillian Murphy)

"Birmingham, 1920s, gangsters… it doesn't immediately sound like an international franchise!" (Steven Knight). But that is what *Peaky Blinders* becomes. Cillian Murphy, Helen McCrory and Paul Anderson lead the cast.

"I mined the memories of relatives regarding the lawless streets of Small Heath, Birmingham, run by gangsters, rocked by turf wars, ruled by the Peaky Blinders… It was about destroying the template of British period drama," says creator Steven Knight.

Peaky Blinders sears across our screens from 12 September. Set in working-class Birmingham, the story arcs from post-World War I and the return of shell-shocked soldiers to the conflict with the IRA, the Russian Revolution, the rise of the British far-right, up to the outbreak of World War II.

The cast is led by Cillian Murphy as charismatic Tommy Shelby in trademark "newsboy" cap and with razor-sharp haircut (both create a fashion wave), along with Helen McCrory (Polly Gray), Paul Anderson (Arthur Shelby), Sophie Rundle (Ada Shelby), and Sam Neill (Inspector Chester Campbell) as the ruthless police chief from Belfast come to impose order on an increasingly lawless city. Later stars include Anya Taylor-Joy, Sam Claflin and Tom Hardy.

The show runs for six series until 2022, gaining a huge following and spawning a festival, video game, clothing line and a beer.

In this year

4 January
The Dumping Ground

2 April
The Great British Sewing Bee

12 September
Peaky Blinders

271

Comedy for All: Mrs Brown's Boys, W1A, Inside No. 9, The Detectorists

"That's nice. Now feck off!"

A classic Mammy catchphrase.

Meet the family, and of course the Mammy (Brendan O'Carroll).

In this year

5 February
Inside No. 9

19 March
W1A

18 September
Following a referendum, Scotland votes to remain part of the UK.

2 October
The Detectorists

23 October
Life Story depicts key moments in animals' lives.

Meet Agnes Brown (Brendan O'Carroll), Irish matriarch, whose favourite pastime is interfering in her six children's lives… The show is a real family affair: O'Carroll's wife, Jennifer Gibney, plays Cathy; his sister Winnie McGoogan plays Eilish; his daughter Fiona plays Maria; his son Danny plays Buster; and his daughter-in-law Amanda Woods plays Betty.

Launched on BBC One on 21 February 2011, the show dates from the late 1990s as a radio comedy, theatre piece, book and Irish TV sitcom. Eventually, BBC Comedy producer Stephen McCrum sees the theatre show and is "knocked sideways by a tsunami of mirth." *Mrs Brown's Boys* proceeds to rack up huge TV audiences, especially around its Christmas Specials.

The show harks back to a more physical, broad-brush comedy, regularly breaking the fourth wall. Occasionally, O'Carroll goes wildly off-script, just to see what the rest of the cast will do.

272

Hugh Bonneville, Jason Watkins and David Westhead lead the perpetually perplexed in *W1A*.

Steve Pemberton and Reece Shearsmith are up to no good *Inside No. 9*.

The Detectorists' world is lovingly created by Mackenzie Crook.

Some TV critics dislike it – too close for comfort to the old drag comedy of Les Dawson and Dick Emery, or else "comedy that makes you vaguely embarrassed to be Irish" (*Irish Independent*). But, as O'Carroll says, on winning best sitcom at the 2013 National Television Awards: "I can only write what makes me laugh, and what makes me laugh is the comedy I grew up on… You hope that the audience agree, and so far, they do."

W1A

John Morton follows up his Olympics satire *Twenty Twelve* (2012) with a mockumentary of the BBC, taking its famous postcode as the show's title. Former Olympic Head of Deliverance Ian Fletcher (Hugh Bonneville) now has a job as the BBC's Head of Values and there is sterling comic support from Jessica Hynes, Jason Watkins, Sarah Parish, Hugh Skinner and Monica Dolan in three series (from 19 March), including a *Lockdown Special*.

Inside No. 9

If you like your comedy darker, then tune in to *Inside No. 9* from 5 February, an anthology of twisted comic tales by Steve Pemberton and Reece Shearsmith (fresh from *The League of Gentlemen*). Flitting deftly from drama to farce, gothic horror to silent movie, the locations range from an echoing mansion, a train's sleeping car and a call centre cubicle, to a karaoke booth, hotel corridor and a police car… Steve Pemberton comments: "There's no other show on TV where people can get to explore this range of different ideas and genres, not only as an actor, but as a writer. So we feel very, very privileged…" *Inside No. 9* wins a clutch of awards, including BAFTA, RTS, Golden Rose and Writers' Guild of Great Britain.

The Detectorists

And finally, take to the fields with friends Lance and Andy (Toby Jones and Mackenzie Crook) as they comb the Essex countryside for hidden treasure, as well as coping with a manipulative ex-wife (Lucy Benjamin), and a girlfriend hoping for better things (Rachel Stirling).

Crook's charming, offbeat BBC Four comedy (he is writer, director and actor) launches on 2 October and was inspired by a *Time Team* episode featuring a couple of detectorists: "They struck me as quite odd characters, very secretive and protective of their hobby, and it just fascinated me…" (Crook).

Wolf Hall

"For them, this is real life. Henry doesn't know he is going to have six wives,
Anne doesn't know her end."

Wolf Hall director Peter Kominsky

Acclaimed stage and screen actor Mark Rylance reinvents Thomas Cromwell as a shape-shifting man of our times.

So we return to the Tudors, but this time it's not to the visible power and pageantry of monarchy, but to the machinations behind it.

Based on Hilary Mantel's bestselling novel, *Wolf Hall* unfolds the story of Thomas Cromwell (Mark Rylance), the son of a blacksmith who becomes Henry VIII's chief minister. Rylance is supported by a superlative cast: Damian Lewis as Henry VIII, Claire Foy as a child-like but capricious Anne Boleyn, Bernard Hill as the Duke of Norfolk, and Anton Lesser as Sir Thomas More.

The adaptation is by Peter Straughan, who grapples with adapting such a long and admired book: "I had the miser's problem of not wanting to lose anything." The director is Peter Kominsky, known for contemporary dramas such as Channel 4's *The Government Inspector*, starring Rylance as David Kelly, the man at the heart of the Iraq War clash between the BBC and the Government. Kominsky gives history a sense of the precarious here and now.

Poldark

"Aidan was our one and only choice."

Writer Debbie Horsfield declares Aidan Turner her perfect pick for *Poldark*.

"Slavery, the Acts of Union in 1800, Acts of Parliament designed to suppress potential revolution, the Napoleonic Wars… It all seems so resonant and relevant to what's going on..."
Poldark adapter Debbie Horsfield.

In this year

21 January
Wolf Hall

8 March
Poldark

29 April
Peter Kay's "will they, won't they" sitcom *Car Share* stars himself and Sian Gibson as work colleagues thrown together by chance.

9 September
Mike Bartlett's thriller *Dr Foster* stars Suranne Jones as a wronged wife.

Broadcasting does one of its famous repeats – *The Forsyte Saga*, *Pride and Prejudice*, *Vanity Fair* and now *Poldark* (8 March). Robin Ellis was a relative unknown when he made a dashing success of the hero of Winston Graham's Cornish saga in 1975; it returns 40 years later with the rougher-edged and charismatic Aidan Turner in the title role.

Captain Ross Poldark returns from fighting in the American War of Independence to find his old love Elizabeth (Heida Reed) married to his weak-willed cousin Francis (Kyle Soller). He marries servant-girl Demelza (Eleanor Tomlinson), battles the devious, money-grabbing Warleggans, and rises to become MP for Cornwall. There's also a role for the original Poldark as the corrupt judge who sends Poldark to gaol.

Adapter Debbie Horsfield, known for Northern-based dramas such as *Cutting It* (2002), amplifies the characters' inner lives – touching on post-traumatic stress disorder, depression and bereavement – as well as conveying the sense of a fragmenting, uncertain Britain. *Poldark* makes a star of Turner, excites debates about male objectivity in his semi-naked scything scene, and leaves Sunday evenings empty for many viewers when it ends, after five series, in 2019.

A Question of Balance: Brexit

"Those who support leaving the EU regard the BBC as the Brussels Broadcasting Corporation. Those who seek to remain, meanwhile, view it as the Brexit Broadcasting Corporation."

Roy Greenslade, *The Guardian*

Across the country, 87 per cent of people follow news of the Referendum result. The story is the most followed since tracking began eight years earlier. A record 53 million global browsers visit BBC News online on the day the result is announced.

Time and again over the decades audiences have come to the BBC for coverage and analysis of major news events, but the EU Referendum ignites new challenges and reveals a highly polarized nation.

The campaign is long and, of course, binary, and the BBC strives for broad balance rather than over-monitored exactitude in its coverage. Even so, the BBC is criticized by both sides for its tit-for-tat news coverage, which creates what one commentator calls a "phoney balance". Critical counts are regularly made of Remain/Leave speakers on flagship programmes such as *Question Time* and the *Today* programme, with neither side content.

Meanwhile, the Internet becomes a crucible for claims and counter claims about the EU and the Referendum. The BBC responds by launching its own Reality Check service. Throughout the campaign – and also across this year's US Presidential Election and the Trump ascendancy where "fake news" becomes a regular mantra – the service challenges and investigates assertions from public figures and weighs them against the available evidence.

The BBC's extensive Referendum coverage includes programmes such as *Brexitcast*, featuring commentary from political correspondents such as Chris Mason, Laura Kuenssberg, Adam Fleming and Katya Adler.

Fleabag

> "If I looked down over the precipice into the chasm below, at the bottom was Fleabag wearing lipstick and looking up at me."
>
> Phoebe Waller-Bridge explains the inspiration for *Fleabag*.

Fleabag (Phoebe Waller-Bridge) gives one of her inimitable "give-away" looks to camera.

In this year

3 January
Tolstoy's epic novel *War & Peace* is dramatized by Andrew Davies and stars James Norton, Lily James and Paul Dano.

21 February
John Le Carré's *The Night Manager* stars Tom Hiddleston, with Hugh Laurie and Tom Hollander playing the chief villains.

June
The BBC's Brexit coverage

21 July
Fleabag

Meet Fleabag. She's not talking to all of us – she's talking to you. So why don't you pop your top off and come right in?

Fleabag is the one-woman Edinburgh Festival Fringe show of 2014 that goes on to become a BBC Three hit (21 July), then a global success, winning its creator Phoebe Waller-Bridge an unheard-of 11 Emmy nominations of which six are winners. "This is just getting ridiculous!" says Waller-Bridge as she wins Outstanding Comedy show of 2019.

Interviewed by Jenni Murray on *Woman's Hour* in 2020, Waller-Bridge talks candidly about where the inspiration for the show came from, about a woman who only felt she was valued because of her sexual desirability: "And I really wanted to write about it in a way that was accessible and funny so people didn't realize that's what it was about until it sneaks up on you."

Asking for trouble

The result is the chaotic, blackly humorous and embarrassingly explicit character of Fleabag (not Waller-Bridge herself, though "I called it *Fleabag*, which is my family nickname, so I was asking for trouble!"). She pushes the boundaries of female identity and gives new meaning to breaking the televisual fourth wall: she takes us deeper and deeper into her personal life, then, unable to escape the camera, starts to regret it because of how much pain is revealed.

Also starring Sian Clifford, Bill Patterson, Olivia Coleman, Hugh Skinner and Andrew Scott (the "hot priest"), *Fleabag* runs for two series and launches Waller-Bridge's career. She is soon spicing up the script of the James Bond film, *No Time to Die* (2021).

Blue Planet II and Climate Change – the Facts

"I've been absolutely astonished at the result that that programme has had. I never imagined there would be quite so many of you who would be inspired to want change."

Sir David Attenborough

The awe-inspiring beauty of the ocean captured in *Blue Planet II*.

A generation on from the BBC Natural History Unit's series *The Blue Planet* (2001), comes *Blue Planet II* on 29 October. Using breakthroughs in marine science and cutting-edge technology, it explores new ocean worlds with an epic visual sweep and reveals the very latest discoveries. Presented once again by Sir David Attenborough, *Blue Planet II* is underpinned by Hans Zimmer's dramatic musical score.

The landmark seven-part series is visually breathtaking, bringing viewers face to face with unexpected new landscapes and includes many filming "firsts" – such as the ingenious tusk fish which uses a rock to break open clam shells, or the incredible hunting teamwork of bottlenose dolphins.

Greta Thunberg: A Year to Change the World, David Attenborough meets the environmental activist in 2021.

The environmental message

However, as we gasp in awe, *Blue Planet II* raises awareness of the immediate threats facing our oceans. So the series simultaneously reveals the devastation human activity has wrought throughout the oceans: the bleached coral reefs and ruinous pollution caused by tons of plastic waste. Voted BAFTA's "must-see moment of the year" is the last episode's heartbreaking sight of a pilot whale carrying in her mouth her dead young, poisoned by plastics in the ocean.

The reaction is momentous: 37 million people in the UK watch the show, and. following the final episode, 62 per cent of those surveyed want to make changes to their daily lives to reduce the impact on our oceans. The BBC also launches Plastics Watch, a pan-BBC initiative promoting the tackling of plastics pollution across the globe.

The climate crisis

Blue Planet II will change the way that the environment is covered by the BBC from this moment on. It leads directly to *Climate Change – The Facts*, which launches on 18 April 2019 and is a specific call to action. Here, David Attenborough takes a stark look at the science surrounding climate change in today's world, detailing the dangers we are already having to deal with – but also the possibilities of prevention and radical political, social and cultural change.

Sir David Attenborough
Opposite page, clockwise from top: Attenborough with armadillo, *Attenborough and Animals*, 1963; close up with humming bird, *Life in Colour*, 2019; close up with giant millipede, *Life in the Undergrowth*, 2005; *Living Planet*, 1984. *This page:* An environmental appeal on the Pyramid Stage at Glastonbury, 2019 (*above*); reviewing his life as Controller of BBC Two in 1993 (*below*).

Dangerous Women: Killing Eve, Happy Valley, Doctor Foster

"You should never tell a psychopath they are a psychopath. It upsets them."

Killing Eve's Villanelle's tell-tale words.

From award-winning producers Sid Gentle Films and BBC America, *Killing Eve* surges into televisual life in 2018, with the first season written by Phoebe Waller-Bridge as an adaptation of Luke Jennings' novellas. Drawn to writing women who say the unsayable and do the undo-able, *Killing Eve* becomes for Waller-Bridge the drama of "transgressive women, friendships, pain". Starring Sandra Oh as British intelligence agent Eve and Jodie Comer as Russian assassin Villanelle, the show plays with and subverts the tropes of traditional spy thrillers and features eye-catching costume design from, among others, Phoebe De Gaye and Sam Perry. *Killing Eve* hits BBC screens on 15 September, going on to four award-winning and critically acclaimed series until 2022, each with a different female lead writer.

Jodie Comer and Sandra Oh are in deadly pursuit in *Killing Eve*.

Sarah Lancashire is after justice in a not-so-*Happy Valley*.

Happy Valley

Meanwhile, in the West Yorkshire setting of the ironically named *Happy Valley*, drug dependency is the problem. Tough, defiant Sergeant Catherine Cawood (Sarah Lancashire) is grappling with her traumatic past in Sally Wainwright's searing, award-winning drama.

Launching on 29 April 2014, it runs across the decade, in 2016 and 2022. We follow the characters living in an extreme circumstances, which have affected them in complex and complicated ways. Eschewing the tension of a classic whodunit, *Happy Valley* is more a study in humanity – focused on Cawood pursuing the man who raped her daughter.

Doctor Foster

And finally Suranne Jones wins every award going as the eponymous star of Mike Bartlett's revenge drama, launched on 9 September 2015 and concluding in 2017. Based loosely on the Ancient Greek legend of Medea, *Doctor Foster* sees Gemma Foster's life unravelling when she discovers her husband (Bertie Carvel) is having an affair (with Jodie Comer's Kate).

Like Wainwright and Waller-Bridge, Bartlett is challenging stereotypical female characters: "I wanted to take some of those misogynist ideas about mad women and witches, and hopefully subvert them. I get upset when people describe Gemma as mad. I don't think she is; she's just very angry."

Dr Gemma Foster (Suranne Jones) with her husband Simon (Bertie Carvel).

Britain Past and Future: A Very English Scandal and Years and Years

> "It's men and power. It's a story that never goes away. Men in power lie, have sex – add a dead dog and you're away."
>
> Russell T Davies

The cast of *Years and Years* bear witness to a dangerously changing world.

On 20 May, erstwhile heart-throb Hugh Grant reinvents himself as a reptilian Jeremy Thorpe in Russell T Davies' *A Very English Scandal*, the story of the closeted gay MP who attempts to kill his lover Norman Scott (Ben Wishaw), and is then saved as the Establishment closes ranks. Adapted from John Preston's book and directed by Stephen Frears, the drama captures "the Englishness of it, the eccentricity of it" (Davies), as well as empathy for both men, trapped where history and society have placed them.

Ben Wishaw and Hugh Grant at the beginning of A *Very English Scandal.*

Years and Years

On 14 May 2019, Davies loops us through time in *Years and Years*. which takes one family in a world of accelerating change and hurls it into the future. "I wanted for many years to write a drama about our civilization sliding… I realized the key to telling that story was through a family." With a cast boasting Emma Thompson, Rory Kinnear, Anne Reid, Jessica Hynes, Russell Tovey, *Years and Years* shows the power of TV drama to touch and implicate us all. As the family matriarch declares in the last episode: "It's the fault of every single person around this table. We did it. We made the world."

RuPaul's Drag Race UK

"There's something dangerous about drag still, and I enjoy that."

Graham Norton, one of the judges of the first series of *RuPaul's Drag Race UK*

Drag has been around at the BBC for a surprisingly long time. In 1938, Douglas Byng was the first man in drag on BBC television – even though it was not strictly permitted then, he circumvented the system by dressing as a woman on the top, a man below, until he finally threw caution to the winds and appeared in full drag.

He was followed by a succession of drag entertainers on TV, from glamorous Danny La Rue and outrageously-bespectacled Edna Everage to caustic Lily Savage. British drag – betraying its anarchic pantomime roots – tended towards the verbally salacious.

Queen of all she surveys – RuPaul in 2018.

Drag royalty

Finally, on 3 October, RuPaul brings her *Drag Race* to the UK, after conquering America. It becomes the biggest hit of BBC Three on iPlayer, receiving over 10 million requests, and creates a different dimension to the BBC, especially among the young. The Vivienne is named RuPaul's Drag Race 1st Superstar. Of the winner, RuPaul says: "The Viv is a perfect example of the 21st-century British queen."

2019 contenders for the *Drag Race UK* title. The ultimate winner is The Vivienne, fourth from left.

I May Destroy You

"The most sublimely unsettling show of the year."

Vulture TV review

"I'm never trying to tell the audience how to feel, what to think or who to judge" – Michaela Coel.

In this year

31 January
The UK leaves the EU.

26 March
Lockdown measures come into force to control the Covid-19 pandemic.

20 April
The *Bitesize* home-learning system helps children while schools are closed.

26 April
Normal People

7 June
I May Destroy You

15 November
Small Axe film series launches.

Self-assured Londoner Arabella is struggling to write her book. Unable to concentrate, she joins friends at a bar – where her drink is spiked and she is raped, blacks out and wakes up next morning with blood trickling down her forehead… Resisting the label of "sexual victim", Arabella takes on the painful, freeing climb to who she could be.

Michaela Coel's drama *I May Destroy You* (8 June) explores across 12 episodes one of the defining themes of our time – the question of sexual consent in contemporary life where "gratification is only an app away" (Coel). It also touches on our sometimes dysfunctional relationships with social media, homophobia, creativity, false allyship, toxic positivity and the commodification of black pain. If that sounds overwhelming, the show is anything but. It has a languid, unfolding quality.

Creative control

Coel goes through 191 drafts to get there, and is given full creative control as its showrunner, director, star and writer. The result is a unique and deeply personal drama, which platforms a very new voice. As Coel questioned in her James MacTaggart Lecture at the Edinburgh TV Festival of 2018: "Is it important that voices used to interruption get the experience of writing something without interference at least once?"

I May Destroy You wins a raft of awards, including Emmys and BAFTAs for outstanding writing and leading actress, as well as widespread critical acclaim.

Small Axe

"A want and a necessity and a need – that's what sparked it."

Steve McQueen on *Small Axe*

Letitia Wright fighting for
what is right in *Mangrove*.

"If you are the big tree, we are the small axe" – that's the Jamaican proverb that inspires artist and filmmaker Steve McQueen to create Small Axe films. Its wording, originally from the Bible then popularized by reggae superstar Bob Marley, is a declaration of defiance against white supremacy.

However, there's also something very personal at the heart of this endeavour, too: McQueen is also seeking to explore his West Indian heritage in London from the 1960s to the 1980s, and to "understand myself, where I came from."

Tributes to Black resilience

Directed by McQueen and written in collaboration with Courttia Newland and Alastair Siddons, *Small Axe* is a five-film anthology. It comprises: *Mangrove*, on the fight for justice for Caribbean restaurant owner Frank Crichlow; *Lovers Rock*, exploring the musical liberation of house-party culture; *Red, White and Blue*, unpacking family tensions across racial injustice in the police force; *Alex Wheatle*, one Black man's journey of self-discovery from orphaned outsider to cultural activist; and *Education*, a tender coming-of-age story and a triumph over inequality.

John Boyega in "The majestic biopic of police officer Leroy Logan" (*The Guardian*), *Red, White and Blue*.

What's remarkable about the *Small Axe* series is its creation of a world, its wrongs and rights, and also its joyous humanity – "The reflection of a fist-pumping revolutionary in the hood of a car, goat curry steaming in a pot, a drop of sweat from a dancer rolling slowly down a gilded wall" (Vox).

In 2020, amid a worldwide reckoning on racial injustice with the killing of George Floyd Jr by a police officer in Minneapolis, this sequence of five such eloquent, beautifully made films could not be more timely. In the words of *Variety* magazine, they are "an auspicious game-changer".

The BBC During the Pandemic

"Tomorrow will be a good day."

Captain Tom Moore's inspiring catchphrase galvanizes the nation in lockdown.

As the national broadcaster, the BBC has always played a crucial role in times of crisis, and the Covid-19 pandemic, which paralyzes the nation this year, is no different.

News is critical – from the expert analysis of health correspondents and behind-the scenes reports at hospitals on the frontline, to intimate radio diaries and personalized data on "Covid in your area". People tune-in in their millions, with requests for news on iPlayer up by over 85 per cent year-on-year in 2020. The BBC News website reaches record figures, with its most-read page ever ("How many cases in your area?") and the highest-ever number of UK weekly browsers. Harmful Covid disinformation is countered by services such as "Reality Check" and "Covid Factchecker".

Lockdown advice wherever you are.

Radio 1's Katie Thistleton inspires some *Lockdown Learning*.

Connell and Marianne (Paul Mescal and Daisy Edgar-Jones) grapple with their emotions in *Normal People*.

Education has always been in the BBC's DNA, and with home-schooling a new reality for most families, the BBC expands its *Bitesize* service into new *Lockdown Learning*, providing students, teachers and often beleaguered parents with immediately available resources. A record-breaking 5.8 million browsers use it in the first week of launch, January 2021. BBC stars are also recruited to add a bit of extra inspiration to the online classroom.

A bit of escapism

The "Entertain" element in the BBC's mission is not forgotten, either. Even though ambitious programme-making becomes well-nigh impossible during the crisis, the BBC continues to supply comedy, music and entertainment shows. iPlayer comes into its own, with more people at home with time to explore the breadth of BBC programmes, past and present. An astonishing 6.1 billion requests to stream programmes is received and top choice is the radio adaptation of Sally Rooney's bestselling 2018 novel *Normal People*: 63.7 million requests. Finally, BBC Local Radio connects intensively to local communities via a campaign called "Make a Difference", reaching out to people in need. A certain proud daughter rings up BBC Three Counties Radio station to let them know that her father is doing laps of his garden to raise money for the NHS. Yes, it's 99-year-old Captain Tom Moore – who will go on to raise over £40 million and receive a knighthood.

2022

As the world comes out of Lockdown, the BBC marks one hundred years of continuous broadcasting by continuing to inform, educate and entertain. The words are the same, but how the BBC realizes its mission continues to evolve, as it has every decade.

INFORM

Over 100 million unique visits to the Ukraine Live page of the BBC News App at the start of the conflict in February 2022.

The war in Ukraine demonstrates the vital need for impartial news to cut through the confusion of claim and counter claim. Chief International Correspondent Lyse Doucet reports from war-torn Ukraine: "Kyiv holds its breath – we all do…"

EDUCATE

Share Your Story tours the UK, working with 400,000 secondary school students to create the next generation of storytellers.

New developments for BBC Bitesize in 2022, as The Regenerators bring green issues to every subject in every classroom, and inspire the next generation to get directly involved.

ENTERTAIN

David Morrissey, Leslie Manville and Robert Glenister in the true-crime thriller *Sherwood*.

Joe Alwyn and Alison Oliver in the relationship drama *Conversations with Friends*, based on the novel by Sally Rooney.

Steven Knight follows *Peaky Blinders* with World War II drama *SAS Rogue Heroes*, starring Alfie Allen, Connor Swindells and Jack O'Connell.

LIVE is back! This year the BBC Outside Broadcast team helps us mark the Queen's Platinum Jubilee with a long weekend of ceremony and celebration, from the Trooping of the Colour to a Party at the Palace.

There's a seat for everyone at Glastonbury, the world's biggest and brightest greenfield pop festival.

Index

Senior Editor: Alastair Dougall
Senior Designer: Nathan Martin
Designers: Robert Perry, Thelma-Jane Robb
Senior Production Controller: Mary Slater
Production Editor: Marc Staples
Managing Editor: Emma Grange
Managing Art Editor: Vicky Short
Publishing Director: Mark Searle

Dorling Kindersley would like to thank:
Archivist James Edwards (BBC Photography Archive);
James Stirling and Daniel Mirzoeff at the BBC;
Rituraj Singh for picture research; Helen Peters for the Index.

First published in Great Britain in 2022 by
Dorling Kindersley Limited
One Embassy Gardens, 8 Viaduct Gardens,
London SW11 7BW

A Penguin Random House Company
Page design copyright © 2022 Dorling Kindersley Limited.

10 9 8 7 6 5 4 3 2 1
001–331499–Sept/2022

For the curious

www.dk.com

AUTHOR'S ACKNOWLEDGEMENTS

With thanks to the invaluable scholarship and research of the BBC's great historians: Asa Briggs, Jean Seaton and David Hendy. Also to my BBC History colleagues John Escolme and Jim McQueen; to James Codd and Hannah Ratford and their team in the BBC Written Archives; to Ralph Montagu, Archive Manager of the *Radio Times*.

PICTURE CREDITS

The publisher would like to thank the following for their kind permission to reproduce their photographs:

(Key: a-above; b-below/bottom; c-centre; f-far; l-left; r-right; t-top)

1 Dorling Kindersley: Clive Streeter / The Science Museum, London. 12 Alamy Stock Photo: John Frost Newspapers (cl). 17 The Society of Authors as the Literary Representative of the Estate of Alfred Noyes: (Quote). 36 Alamy Stock Photo: De Luan (cla). 46 Alamy Stock Photo: Trinity Mirror / Mirrorpix (cla). 82 Alamy Stock Photo: BBC / Moviestore Collection Ltd (c). 83 Alamy Stock Photo: Moviestore Collection Ltd (tl). 126 Alamy Stock Photo: Moviestore Collection Ltd (cr). 137 Coolabi Productions Limited: Clangers © 2022 Coolabi Productions Limited, Smallfilms Limited and Peter Firmin (bl). 139 Alamy Stock Photo: Ian Bottle (br); Moviestore Collection Ltd (tr). 178-179 Alamy Stock Photo: Trinity Mirror / Mirrorpix. 199 Alamy Stock Photo: PA Images (b). 216 Alamy Stock Photo: Moviestore Collection Ltd (clb); Photo 12 / Photo: Giles Keyte; BBC Films / Working Title Films (cla). 217 © ear for eye Ltd, British Broadcasting Corporation and The British Film Institute 2021 / Written & Directed by debbie tucker green / Produced by Fiona Lamptey. 226 Lookout Point: ASB image (bl). 243 Brian Ritchie | © Talkback / QI Ltd: (bl). 248 Boundless/BBC: (bl). 255 Hat Trick: (br). 256 Aardman Animations Ltd 2008: 'A Matter of Loaf & Death © 2008 Wallace & Gromit Ltd (cra). 257 BBC Children: Lion TV (cla). 263 Wall to Wall Media Limited: (tl). 268 Neal Street Productions: (cra). 269 World Productions: (cla, bl). 271 Caryn Mandabach Productions Ltd: (ca). 274 BBC Picture Archives: Jeff Overs (c). 275 Mammoth Screen: (cra). 277 Two Brothers Pictures: (cla). 282 Sid Gentle: (bl). 283 Drama Republic: (br). Red Production Company: (tl). 284 Blueprint Pictures: (cr). Red Production Company: (bc). 286 Various Artists Limited: Falkna Productions (cla). 287 Turbine Studios: (cla, bl). 289 Element Pictures: © EP Normal People Limited / Enda Bowe, 2020 (bl)

Cover images: Front: **Dorling Kindersley:** Clive Streeter / The Science Museum, London

All other images copyright © BBC.

QUOTES

Page 10: Lyrics from "I'll B.B.C-in' You": Marion Harris and Reginald Montgomery, published in 1935 by Lawrence Wright of London.

Page 17: Extract from "The Dane Tree": The Society of Authors as the Literary Representative of the Estate of Alfred Noyes.

Page 44: Extract from "Television", lyrics by James Dyrenforth, music by Kenneth Leslie-Smith.

Page 56: Extract from "Little Brown Girl" copyright Una Marson.

Page 66: Lyrics from "We Want Muffin (Muffin the Mule)" by Annette Mills. Copyright 1948 Sterling Music (an affiliate of Chappell & Co).

Page 88: Quote from *Under Milk Wood* by Dylan Thomas by permission of © The Dylan Thomas Trust.

Page 194: Lyrics from "The Ballad of Barry and Freda" by Victoria Wood, used by permission of the Victoria Wood Estate.